Studies in Logic
Volume 32

Foundations of the Formal Sciences VII
Bringing together Philosophy and Sociology of Science

Volume 27
Inconsistent Geometry
Chris Mortensen

Volume 28
Passed Over in Silence
Jaap van der Does

Volume 29
Logic and Philosophy Today, Volume 1
Johan van Benthem and Amitabha Gupta, eds

Volume 30
Logic and Philosophy Today, Volume 2
Johan van Benthem and Amitabha Gupta, eds

Volume 31
Nonmonotonic Reasoning. Essays Celebrating its 30th Anniversary
Gerhard Brewka, Victor W. Marek and Miroslaw Truszczynski, eds.

Volume 32
Foundations of the Formal Sciences VII. Bringing together Philosophy and Sociology of Science
Karen François, Benedikt Löwe, Thomas Müller and Bart Van Kerkhove, eds.

Volume 33
Conductive Argument. An Overlooked Type of Defeasible Reasoning
J. Anthony Blair and Ralph H. Johnson, eds.

Volume 34
Set Theory
Kenneth Kunen

Volume 35
Logic is not Mathematical
Hartley Slater

Volume 36
Understanding Vagueness. Logical, Philosophical and Linguistic Perspectives
Petr Cintula, Christian G. Fermüller, Lluís Godo and Petr Hájek, eds.

Volume 37
Handbook of Mathematical Fuzzy Logic. Volume 1
Petr Cintula, Petr Hájek and Carles Noguera, eds.

Volume 38
Handbook of Mathematical Fuzzy Logic. Volume 2
Petr Cintula, Petr Hájek and Carles Noguera, eds.

Studies in Logic Series Editor
Dov Gabbay   dov.gabbay@kcl.ac.uk

Foundations of the Formal Sciences VII

Bringing together Philosophy and Sociology of Science

Edited by

Karen François,
Benedikt Löwe,
Thomas Müller
and
Bart Van Kerkhove

© Individual author and College Publications 2011.
All rights reserved.

ISBN 978-1-84890-049-3

College Publications
Scientific Director: Dov Gabbay
Managing Director: Jane Spurr
Department of Informatics
King's College London, Strand, London WC2R 2LS, UK

http://www.collegepublications.co.uk

Original cover design by Orchid Creative   www.orchidcreative.co.uk
This cover produced by Laraine Welch
Printed by Lightning Source, Milton Keynes, UK

All rights reserved. No part of this publication may be reproduced, stored in a retrieval system or transmitted in any form, or by any means, electronic, mechanical, photocopying, recording or otherwise without prior permission, in writing, from the publisher.

Contents

Preface vii

Knowledge, the context distinction and its impact on the relation between philosophy and sociology of science
Sabine Ammon 1

Science as socially distributed cognition: Bridging philosophy and sociology of science
Matthew J. Brown 17

Freud's unintended institutional facts
Filip Buekens & Maarten Boudry 33

Looking for Busy Beavers. A socio-philosophical study of a computer-assisted proof
Liesbeth De Mol 61

Sources for myths about mathematics
Christian Greiffenhagen & Wes Sharrock 91

On the curious historical coincidence of algebra and double-entry bookkeeping
Albrecht Heeffer 111

Economic calculation. Frameworks and performances
Herbert Kalthoff 133

Demystification of early Latour
Jouni-Matti Kuukkanen 161

Albert Lautman: Dialectics in mathematics
Brendan Larvor 185

On the philosophical talk of scientists
Hauke Riesch 205

Career paths in mathematics: A comparison between women and men
Renate Tobies 229

Alternative claims to the discovery of modern logic: Coincidences and diversification
Paul Ziche 243

Preface

Both philosophy and sociology of science aim at understanding the workings of scientific endeavour. Despite their different emphasis and possibly methodology, they deal with the same subject matter. And yet, they seem to be worlds apart. The 20th century has seen Western philosophy being divided in two big currents: a newly arising Anglo-Saxon or analytic one, and a traditional European or continental one. There is very little communication between these two areas of philosophy. A precise meta-philosophical definition of these two currents is difficult, if not impossible, and yet the differences in approach are striking. This seems to suggest that what unites or divorces philosophers in and across both traditions be better thought of in terms of family-resemblances instead of radical opposites. The study of sociology of science is strongly influenced by the continental tradition, whereas the philosophy of science has been the near exclusive playground for analytic approaches.

Second, there is the central issue of the so-called *science wars*: the question of the the proper relationship between humanities and natural sciences. To wit, philosophy of science has been (and is) predominantly the philosophy of natural science, while the sociology of science is almost exclusively conducted from within humanity faculties that are fairly remote to the actual practices of the exact sciences they purport to describe.

In the early days, the sociology of science explicitly set its task as being complementary to that of philosophy, but current sociology of science focuses on social organization, epistemic content and cultural aspects of science, breaking down the barrier respected by their ancestors, resulting in an approach seemingly incompatible and openly at odds with that of philosophy of science.

The good news is that, in the course of the last few decades, steps have been taken towards a (partial) reconciliation. We see our conference FotFS VII as part of this process, bringing sociological aspects into philosophy of science and philosophical aspects into sociology of science, by bringing together researchers from both areas.

Given that the conference series is concerned with the *formal sciences*, we have a certain, but non-exclusive focus on the role of mathematics as one of the sciences covered by philosophy and sociology. It is therefore fitting that the network PhiMSAMP (*Philosophy of Mathematics: Sociological Aspects and Mathematical Practice*) is supporting our conference.

The FotFS conference series.

The conference was the seventh in the conference series on the "Foundations of the Formal Sciences" (FotFS), a series of interdisciplinary conferences in mathematics, philosophy, computer science and linguistics. The main goal is to reestablish the traditionally strong links between these areas of research, some of which have been lost in the past decades. FotFS started in 1999 as a small German workshop in Berlin. Its defining features were present from the very first meeting onwards: a strong interdisciplinary spirit, a focus on technical talks that nevertheless reach out to researchers from other communities, and a (non-exclusive) focus on young researchers.

After its inaugural meeting in Berlin, FotFS was funded as a "PhD Euro-Conference" by the European Community, the DFG (Deutsche Forschungsgemeinschaft) and the BIGS (Bonn International Graduate School). Each of the meetings has a distinctive topic specifying some part of the foundations of formal sciences to be investigated in an interdisciplinary way. FotFS II dealt with Applications of Mathematical Logic in Philosophy and Linguistics, and FotFS III with "Complexity in Mathematics and Computer Science". FotFS IV was a meta-conference discussing the topic of the series under the header "The History of the Concept of the Formal Sciences". The conference FotFS V on "Infinite Games" was held in Bonn, and the immediately predecessor of our conference, FotFS VI was held in Amsterdam under the title "Reasoning about Probabilities and Probabilistic Reasoning".

Foundations of the Formal Sciences VII (FotFS VII) took place from 21 to 24 October 2008 in the *Rubensauditorium* of the *Koninklijke Vlaamse Academie van België voor Wetenschappen en Kunsten* (Royal Flemish Academy of Science and Arts). The conference was organized by the *Centre for Logic and Philosophy of Science* of the *Vrije Universiteit Brussel* jointly with the *Wissenschaftliches Netzwerk* PhiMSAMP, and its Scientific and Organizing Committee consisted of Patrick Allo (Brussels), Benedikt Löwe (Amsterdam), Karen François (Brussels), Thomas Müller (Utrecht), Jean-Paul Van Bendegem (Brussels), and Bart van Kerkhove (Brussels). The conference was co-located with the workshop PhiMSAMP-4 (25–26 October 2008).

The following list documents all talks invited or accepted for presentation at FotFS VII (including those that had to be cancelled due to various reasons):

Sabine Ammon. *Reconstruction versus Construction: The Context Distinction and its Impact for Philosophy and Sociology of Science.*

Alexandre Borovik. *Science Wars: a Time for a Truce.*

Matthew Brown. *Science as Socially Distributed Cognition: Bridging Philosophy and the Sociology of Science.*

Filip Buekens and Maarten Boudry. *Institutional Facts or Social Constructions? A Searlean Reconstruction of Psychoanalytic Facts.*

Bernd Buldt. *Husserl's Theory of Objectivity.*

Jessica Carter. *The use of diagrams in mathematical reasoning.*

Helen De Cruz and Johan De Smedt. *Cognitive and cultural factors influence the spread of mathematical concepts: The case of zero.*

Liesbeth De Mol. *On the use and (interactive) role of computers in computer-assisted proofs.*

Till Düppe. *Listening to the Music of Reason: Nicolas Bourbaki and the Phenomenology of Mathematical Experience.*

Karen François and Bart Van Kerkhove. *Ethnomathematics as an implicit philosophy of mathematics (education).*

Steve Fuller. *On formal/mathematical notions underlying modern progressive thinking.*

Norma B. Goethe. *Modes of Representation and Working Tools in Leibniz?s Intellectual Workshop.* (CANCELLED)

Christian Greiffenhagen. *Formal versus Practical? Opposition to formalism in the sociology of science and mathematics.*

Ari Gross. *Feynman Diagrams and Visual Reasoning.*

Albrecht Heeffer. *On the curious historical coincidence of algebra and double-entry bookkeeping.*

Herbert Kalthoff. *Doing/Undoing Calculation: Sociological Insights from Risk Management.*

Jouni-Matti Kuukkanen. *Inevitable or Contingent History of Science? Re-specifying the Difference between Scientific Realism and the Sociology of Scientific Knowledge.*

Brendan Larvor. *Mathematics, Phenomenology and Social Cognition.*

Frank Linhard. *Formal and non-formal approaches to the notion of "Risk"—a historical perspective.*

Geerdt Magiels. *Towards an Ecological Understanding of Science: the case of the discovery of photosynthesis.*

James McAllister. *Symmetries and asymmetries in science studies.*

Eric Oberheim. *Incommensurability and Reconciliation.*

Hauke Riesch. *On the philosophical talk of scientists.*

Georg Schiemer. *Carnap's early semantics: models, isomorphism and categoricity.*

Marc Staudacher. *Around 20 notions of conventions and still no clue?* (CANCELLED)

Ana Teixeira-Pinto. *The philosophical concepts of autonomy and automatism and the process of mathematization of mechanics.*

Renate Tobies. *Career Paths in Mathematics: Women and Men by Comparison.*

Roy Wagner. *Mathematical variables as indigenous concepts.*

Paul Ziche. *The multiple discovery of logic around 1900: Interactions between Philosophy, Mathematics and the Cultural world.*

The conference FotFS VII was financially supported by the *Deutsche Forschungsgemeinschaft* (DFG) via the *Wissenschaftliches Netzwerk* PhiMSAMP (MU1816/5-1) and the *Nationaal Centrum voor Navorsingen in de Logica / Centre National de Recherche de Logique* (NCNL/CNRL) in Belgium.

This volume is the proceedings volume of FotFS VII for which we received 18 submissions. Twenty-six expert referees helped us to select the twelve papers published in this volume. Brendan Larvor's paper is the English version of a paper published in French as *Albert Lautman, ou la dialectique dans les mathématiques* in the journal *Philosophiques* (37:1 (2010), pp. 75–94), and we would like to thank *Librairie philosophique J Vrin* and the editor of the journal *Philosophiques*, Professor Christian Nadeau, for their kind permission to include it in this volume. We would also like to thank *Springer-Verlag* for their permission to reprint two illustrations from a 1934 paper published in the journal *Erkenntnis* in Paul Ziche's article in this volume.

The typesetting and printing of this volume was funded by the DFG (MU1816/5-1); in addition, we should like to acknowledge the help of David Fiske (for correcting the English of some of the papers) and Edgar Andrade (for his bibliographical and typesetting support).

June 2011, Brussels, Hamburg, and Utrecht

K. F. B. L. T. M. B. V. K.

The conference photo.

François, K., Löwe, B., Müller, T., Van Kerkhove, B., editors,
Foundations of the Formal Sciences VII
Bringing together Philosophy and Sociology of Science

Knowledge, the context distinction and its impact on the relation between philosophy and sociology of science

SABINE AMMON

Nationaler Forschungsschwerpunkt eikones "Bildkritik. Macht und Bedeutung der Bilder",
Rheinsprung 11, 4051 Basel, Switzerland
E-mail: Sabine.Ammon@unibas.ch

1 Introduction

The relation between philosophy and sociology of science can be characterized by an impressive gap, which seemed to be unbridgeable for several decades. This might be astonishing as both disciplines deal with the same subject matter and both disciplines try to understand what science is. Indisputably, there are various overlaps between the two disciplines and one would expect a lively, interdisciplinary discussion. However, the opposite is the case. To this day, there are persistent obstacles, which hamper a fruitful exchange. Bringing together philosophy and sociology of science, but also history of science, to a joint interdisciplinary field remains a desideratum. Since the early 1970s, there has been progress in establishing science studies, which include historical, sociological, ethnological, psychological, cultural, political and economical aspects. However, to this day there have been serious difficulties to integrate mainstream philosophy of science into this dialogue. The historians of science Schickore and Steinle state that

> exchanges between philosophy of science, history of science and science studies have been rather sparse; in fact, the disciplines have drifted further and further apart. (Schickore and Steinle, 2006a, p. ix)

Consequently, Weingart, a sociologist of science, concludes that

> the differences between the formal orientation of theory of science and the empirical orientation of sociology of science are too large,

Received by the editors: 24 January 2009; 18 November 2010; 5 April 2011.
Accepted for publication: 14 April 2011.

and the institutional interests of the developed fields relate to these differences. (Weingart, 2003, p. 12; translation by the author)[1]

These quotations show us a concise, but rather depressing diagnosis of the status quo. However, we should not leave it at that diagnosis, as the loss for both, philosophy and sociology of science, is remarkable. A pressing, present-day problem, which demands for a joint treatment, is the question of knowledge in its relation to science and society. Especially in the German-speaking context, a virulent debate in social science on the effects of the knowledge society is taking place.[2] Knowledge has become a key concept for analysing especially the fields of education, research, science and politics. However, despite its prominent role the notion of "knowledge" usually remains diffuse and vague. Philosophy of science could be the ideal partner for sociology of science to tackle this problem, as clarifying concepts is a genuine philosophical endeavour. Nevertheless, philosophical treatises on theory of knowledge rarely enter into writings of sociology of science or sociology of knowledge. Why is this so? At first glance, some causes are ready at hand. The debate in Epistemology and Philosophy of Science on theory of knowledge[3] is confusing (not only for insiders) as we find a great variety of positions. The majority of debates centre on the notion of knowledge as "justified true belief" in search of necessary and sufficient conditions. The price of this approach is a very narrow notion limited to propositional knowledge (e.g., Steup, 2008; Baumann, 2002, esp. pp. 40 ff.). Many occurrences of knowledge like knowing-how, knowledge by acquaintance or the difference between manifestations of knowledge and personal knowledge cannot be covered by these definitions,[4] which result in fragmented conceptions of knowledge. Therefore, most of the analysis done in theory of knowledge seems unsuitable for use in social studies. By discussing the relation between society, individuals, science and knowledge, sociology

---

[1] German original: "Zu groß sind die Unterschiede zwischen der formalen Orientierung der Wissenschaftstheorie und der empirischen Orientierung der Wissenschaftssoziologie, und mit diesen Unterschieden verbinden sich die institutionellen Interessen der gewachsenen Fächer."

[2] For an overview, cf., e.g., Weingart (2003, pp. 127–141), Knoblauch (2005, pp. 255–284), Maasen (2009, pp. 77–83).

[3] Bringing together Philosophy of Science and Epistemology means bridging another far-reaching gap lurking in theoretical philosophy, which should be mentioned here but will not be discussed further. With the beginning of the 20th century, Philosophy of Science and Epistemology started to drift apart and developed into different fields. Therefore, the discussion of problems of knowledge—which is crucial to both fields—has led to different approaches, which remain mostly unrelated. A proposal for how to solve many problems concerning theory of knowledge by bringing together Philosophy of Science and Epistemology can be found in Ammon (2009), which develops a procedural notion of knowledge.

[4] As, e.g., argued in Gottschalk-Mazouz (2007, pp. 22 ff).

of science needs a comprehensive theory of knowledge, which is able to explain how different forms of knowledge interact, how knowledge changes in history or how knowledge depends on cultural and societal influences.

Given these facts, the need for an integrative discussion of philosophy and sociology of knowledge for opening traditional lines of argumentation and analysis is clear. This is astonishing as one can observe an increasing mutual interest during the last two decades.[5] In epistemology, the field of social epistemology emerged, which acknowledges and investigates the social dimensions of knowledge with influential writings by Steve Fuller (1988), Alvin Goldman (1999) or Helen Longino (2002)—to mention only some protagonists. In philosophy of science, we could for example observe a re-evaluation of the discovery process by the "friends of discovery"[6] and an evolving literature on the epistemology of experimentation when we look at texts like Nickles (1980a,b), Hacking (1983), Franklin (1986) or Rheinberger (1997). However, in spite of this body of literature, mainstream debates appear rather reluctant when it comes to a renewal of discourse concerning the problem of knowledge. There seem to be obstacles deeply anchored in the disciplinary discourse, going beyond different methodological approaches.

Therefore, the aim of this article is to single out one of these obstacles in the field of theory of knowledge which must be seen in close relation to the context distinction. The establishment of the context distinction—as the separation into a context of discovery and a context of justification—goes back to the beginning of the 20th century and leads to a specific handling of problems in theory of knowledge.[7] Using the example of Carnap's *Aufbau* (1928), it is possible to show a reconstructional attitude, which is not only paradigmatic for the traditional theory of knowledge in philosophy[8] but also contributes to the isolation of epistemological from sociological points of view to this day. However, the discussion of Carnap's approach can also fuel the search for a solution for how to overcome this obstacle. The writings of Nelson Goodman, who has developed the epistemic concept of the *Aufbau* further, can serve as a point of departure, and allow us to contrast the *reconstructional* approach to a *constructional* emphasis. Based

---

[5] The conference FotFS VII at which this paper was presented can be seen in this line as well; cf. also Ammon et al. (2007a).

[6] Cf. Sintonen and Kiikeri (2004, pp. 214 ff.); according to Schickore and Steinle, the expression "friends of discovery" goes back to Ronald Giere (Schickore and Steinle, 2006a, p. viii).

[7] For a detailed overview of the different usages of the notion "context distinction", cf. Hoyningen-Huene (1987, 2006).

[8] To be more precise, Carnap's *Aufbau* exemplifies a common approach to theory of knowledge in Philosophy of Science: The *Aufbau* develops a notion of knowledge originating from the notion of system. Herein lies an important difference: Theories of knowledge in epistemology usually originate in discussing the conditions of knowledge of a person (as it occurs in the standard expression "S knows that p").

on a dynamic plurality of knowledge systems, a shift in theory of knowledge becomes possible, allowing us to adjust the relation of philosophy and sociology of science.

2 The relevance of the context distinction for the relation of philosophy and sociology of science

The separation into context of discovery and context of justification became a paradigm for philosophy of science in the 20th century. Although not explicitly addressed, the context distinction is up to now highly influential for the disciplinary attitude. Schickore and Steinle (2006a, p. vii) observe that

> [the context distinction] still informs our conception of the content, domain, and goals of philosophy of science. The fact that new developments in philosophy of experimentation and history and sociology of science have been marginalized by traditional scholarship in philosophy indicates that the context distinction still pervades philosophical thinking about science.

The context distinction draws a demarcation line between anything that relates to the process of discovering insights and a justification of the knowledge gained in these processes. Nevertheless, it is more than a line between different research areas as only the latter is usually of interest for philosophy of science. According to Schickore and Steinle (2006a, p. vii), one can expect two effects of this demarcation line. On the one hand, it limits the scope of philosophy of science and hence, one can characterize its domain and fix its methods. On the other hand, it is possible to adjust the relation of philosophy of science to other disciplines. As a consequence, philosophy of science becomes a self-assured discipline, which deals with the results of science on a meta-level. Rival disciplines, which also deal with science as a subject matter, but concentrate on the process of discovery, do not need to be considered.

Although there are earlier manifestations of the context distinction, it rose to fame through Reichenbach's "Experience and Prediction" (1966). Reichenbach claims that epistemology should limit itself to "rational reconstruction" of epistemic processes. What is of interest is a logically based reformulation of the insights achieved in epistemic processes, not the thought proces itself that has led to the result. Reichenbach stresses the point that there are differences in the form "in which thinking processes are communicated to other persons" and the form "in which they are subjectively performed" (Reichenbach, 1966, p. 6). As an example, he discusses the results of the work of a mathematician or of a physicist. There is, he states, "the well-known difference between the thinker's way of finding this theorem and his way of presenting it before a public" (Reichenbach, 1966, p. 6). And he continues:

> I shall introduce the terms *context of discovery* and *context of justification* to mark this distinction. Then we have to say that epistemology is only occupied in constructing the context of justification. (Reichenbach, 1966, pp. 6 f.)

Due to the difference in ways of discovery and ways of justification, Reichenbach concludes that it is possible to investigate the context of justification in isolation. The context of discovery that mingles with psychological and social aspects can be left aside. By this, it was possible to separate epistemology from psychology, which was a strong concern in that era (Richardson, 2006, p. 41). Nevertheless, not only psychological, also social and political considerations could be excluded—a motivation targeted at Otto Neurath, Philipp Frank, and the left wing of the Vienna Circle, as Don Howard (2006) argues. Finally yet importantly, Reichenbach intended to establish "scientific philosophy" as a proper discipline and to guarantee the autonomy of epistemology by his argumentation (Schiemann, 2006, pp. 23 f.).

In fact, the distinction became not only extremely influential for the foundation of the discipline of philosophy of science; it also set the standard for handling systematic problems in theories of knowledge. The *justification* of knowledge through epistemological analysis became a major focus. The genesis of knowledge can be neglected—only those aspects that relate to the results of the processes are of epistemic interest. This, in turn, leads to the establishment of the gap between the disciplines mentioned above: On one side of the gap, historians and sociologists concentrate on questions related to discovery and usually ignore the ongoing discussions in philosophy.

> [H]istorians and sociologists have largely ignored epistemological concepts and debates in their historical studies and thick descriptions of specific scientific episodes. (Schickore and Steinle, 2006a, pp. ix–x).

On the other side of the gap, the exclusive concentration on the results of epistemic processes in philosophy of science leads to a stereotypical notion of knowledge, which is far from reality.

> [P]hilosophers working in the analytic tradition continue to exclude historical as well as sociological and psychological studies of science from philosophical reflection. [...] Many [...] simply presuppose that investigations of the material culture, the historical changes, and the cultural and social environments of science lack epistemological significance. (Schickore and Steinle, 2006a, p. x).

Indisputably, there is a long tradition of mutual neglect in both disciplines. As argued before, it is important to bridge this gap; but why are the communities still so reluctant to build this bridge? The answer seems simple. As long as theorists of knowledge can argue, as Schickore and Steinle have pointed out, that the context of discovery lacks epistemological significance, there is no need for building a bridge. The crucial question is therefore, why should discovery matter for epistemology? The situation changes if it is possible to show that a theory of knowledge exclusively based on the context of justification turns out to be inconsistent. In fact, many problems haunt traditional theory of knowledge. If it is possible to show that many of those problems vanish when the context of discovery is reconsidered for justification, a strong argument against the context distinction would be found.

3   A reconstruction of knowledge

In order to gain further clarification, it helps to go back to the roots of the problem. The method of *rational reconstruction*, to which Reichenbach is referring, originally stems from (Carnap, 1928). We do not only owe the terminology of the method to him, but also a first paradigmatic formulation of the context distinction.

> It must be possible to give a rational foundation for each scientific thesis, but this does not mean that such a thesis must always be discovered rationally, that is, through an exercise of the understanding alone. After all, the basic orientation and the direction of interests are not the result of deliberation, but are determined by emotions, drives, disposition and general living conditions. ... The *justification*, however, has to take place before the forum of the understanding; here we must not refer to our intuition or emotional needs. (Carnap, 1928, p. xvii)

Carnap's aim is to ground knowledge on a firm basis, separating it from findings that are wrong, senseless or meaningless: that is to say, no knowledge at all. The thrust is twofold: on the one hand, knowledge should be characterized; on the other hand, a critique of metaphysics should show what cannot qualify as knowledge. The intention is to unmask so-called pseudoproblems and to free epistemology from pure speculation. Carnap talks of purification and cleaning; he stresses the point that it is important to lock out feeling, instinct, disposition or circumstances of one's life. They might be relevant for the genesis of knowledge, but not for the justification of knowledge.

In order to realize his aim, Carnap goes on to collect all findings that exist at a certain time in a unifying system. The focus is language-based: knowledge is characterized as coming in statements that relate to each other. Therefore, *rational reconstruction* is a reconstruction in language by using means of modern logic. By choosing basic concepts, it is possible to develop a so-called family tree of concepts. Within the "constructional system"[9], it is possible to derive any meaningful concept from the basis. Carnap's way of approaching the problem promises a breakthrough for theory of knowledge. Any meaningful statement can be reformulated—and the reformulation uses only precisely determined concepts. A complex system emerges which can integrate not only any currently available knowledge, but also any future knowledge. It is possible to draw an unambiguous line between statements that represent knowledge and statements that are false or meaningless. Knowledge claims which can be reformulated within the system count as knowledge. On the other hand, those which cannot, are revealed as metaphysics as one does not succeed in tracing them back within the constructional system.

Obviously, it is Carnap's procedure to build constructional systems, which implements the context distinction. *Rational reconstruction* can be described more precisely according to the terms involved. On the one hand, the method used is a rational procedure. That is, it follows certain criteria considered as rational. It simplifies, idealizes, systematizes and it follows the requirements of logic. On the other hand, the method comes as a reconstruction: Carnap chooses a retrospective perspective to systematize all available knowledge. To be fair, Carnap gives priority to epistemological primacy in this process of reconstructing. However, this is not mandatory, and Carnap stresses the point that the genesis of knowledge is not to be considered; what matters is its justification.

In a certain respect, Carnap's way of handling the problem of knowledge is paradigmatic. His *rational reconstruction* reveals a solution for a classical problem within the new framework provided by the linguistic turn. Knowledge becomes a system of statements that evolve from some fundamental units and principles. These circumstances lead to a notion of knowledge which can be characterized as follows: (a) knowledge can be reformulated within a system of statements; the focus is therefore on propositional knowledge; (b) the aim is to draw a strict line between what proves to be knowledge and that what is senseless, false and therefore not knowledge; this can be attained by an appropriate procedure of justification; (c) knowledge turns out to be true and certain as it has passed a rigorous procedure of justification; (d) therefore, knowledge is timeless and objective; as a consequence, it accumulates.

---

[9]In order to find an equivalent for the original German term "Konstitutionsystem", the English translation of Carnap's *Aufbau* from 1967 by Rolf A. George (Carnap, 2003) uses the expression "Constructional System"; and already Goodman (1951) introduced the terms "construction" and "constructional" when discussing the *Aufbau*. Therefore, the notion "Constructional System" is widely used in the secondary literature on the *Aufbau*. However, this translation is rather misleading as Carnap names the method employed for system-building "rational reconstruction" which differs significantly from a construction. Nevertheless, for sake of consistency, this article uses the established English terminology.

Although ways and means vary in the ongoing discussion, it is the implicit reconstructional attitude within these characteristic features that remains highly influential and sets the direction for theory of knowledge to this day. In this line of argumentation, it is possible to dismiss any criticism which refers to the limited scope of the notion of knowledge or to the lack of social and historical aspects as it remains external. Its internal setting is consequential and conclusive. As the *reconstructional perspective* on knowledge only works with the *results* of the process of knowing, it causes ahistoricity and the ignorance of social questions in theory of knowledge. For the setting of the system of knowledge, the process of discovery does not need to be considered. By this, rational reconstruction not only reinforces the context distinction. The reconstructional perspective also leads to an immunization of traditional theory of knowledge.

At first sight, the only problem for traditional theory of knowledge is to give an account of the justification of knowledge. However, this problem is not easily solved. The solution of Carnap's *Aufbau* is not satisfying in the long run. Problems of foundations, testing and verification make clear that it becomes impossible to attain the certainty wanted. For example, sentences are embedded in a holistic setting of other sentences; they are involved in dynamic practices and use. The process of drawing a line between knowledge and non-knowledge turns out to be less strict and certain. Especially language, which seemed to be such an appropriate means, creates many problems. Language is no neutral means for reformulating statements of knowledge: on the contrary, it turns out to be full of assumptions. Language itself represents knowledge that captures the world in its notions and structures. In the train of uncovering the prerequisites of language and its relation to the world, modes of acquisition, translation and construction come into focus. In order to ground knowledge on a solid basis, it seems essential to clarify these prerequisites first.

4 Constructing understanding

Not astonishingly, there are many attempts to resolve the deficiencies of traditional approaches in theory of knowledge. An especially fruitful example is represented by the writings of Nelson Goodman, as his ideas evolve through a critical analysis of Carnap's *Aufbau*. Goodman's own treatment of constructional systems lays the foundation for a revised epistemology in his late writings, especially in "Ways of Worldmaking" (Goodman, 1978) and "Reconceptions in Philosophy and other Arts and Sciences" (Goodman and Elgin, 1988). In these works, Goodman demonstrates the shortcomings of classical theory of knowledge that led his way from reconstructional systems to constructional systems and from knowledge to understanding.

An important insight precedes Goodman's epistemological shift. It is the fact that we can find cognition in manifold processes: when we analyse and synthesize, when we observe and create, when we discern, order, weigh or structure. The "cognitive work" is done

> with achieving a firmer and more comprehensive grasp, removing anomalies, making significant discriminations and connections, gaining new insights (Goodman and Elgin, 1988, p. 158).

The notion of understanding summarizes all these processes. What they share is a novel epistemic focus, a change in current structures and systematizations that relates to epistemic merit. According to Goodman's approach, cognitive functioning relates to symbols. With this, he opens a broad range for epistemology. From this point onwards, not only language-based systems are worth considering, but also any symbol system found in science, in our every-day world or in the arts.

Goodman describes the processes of understanding as ways of constructing which are equivalent to the usage of symbol systems. However, it is not just any construction that is the sought-after cognitive activity. It is important to single out those ways of constructing which lead to epistemic success. Therefore, the search for rightness of construction is of primary concern in the novel epistemic focus. The model of justification developed by Goodman (1954) became famous under the heading of *reflective equilibrium* (cf. also Rawls, 1971; Elgin, 1996, pp. 106 ff.; Elgin, 1999, pp. 49 ff). Rightness is to be found with the help of a reflective equilibrium between existing structures and new changes.

> Since rightness is not confined to those symbols that state or describe or depict, the fitting here is not a fitting *onto*—not a correspondence or matching or mirroring of independent Reality—but a fitting *into* a context or discourse or standing complex of other symbols. (Goodman and Elgin, 1988, p. 158)

New symbols are brought in interplay with existing symbol structures in an active and creative process. Changes are made, often on both sides. If the procedure succeeds, if it is possible to attain a new structuring, we have gained a new understanding.

To appreciate the implied dynamics it is important to notice that the process never finds a definite ending. Goodman and Elgin describe the processes that lead to a reflective equilibrium, such as searching, testing and trying out. Novel circumstances, alterations in assessment, or new findings put the established usage in question, which, in turn, leads to a reassessment of the reflective equilibrium. Therefore, those processes never really come to an end. Granted, there are well-entrenched constellations that hardly lead to adjustments—but in principle, any element of established systematizations can become part of the equilibration. Additionally, the criteria of rightness can change. Being part of the processes, they are implicitly tested while probing and examining the fitting of a novel structure. If necessary, the changes not only affect symbols and systems, but also the criteria themselves. Thereby, gaining insights turns out to be an ongoing process of transformation.

Goodman and Elgin (1988, pp. 161 f.) call both the process of constructing and its results "understanding" in order to stress the novel focus in epistemology. However, the chosen terminology conceals the crucial difference between the process and its results. Therefore, in the following argumentation, "understanding" will be restricted to the cognitive process, whereas "knowledge" will used to describe the results of the process of understanding.[10] Consequently, the novel focus in epistemology leads to a revised notion of knowledge which can be characterized as follows: (a) the scope of knowledge is wide-ranging as knowledge is no longer limited to systems of statements but embraces all symbol systems; (b) the rightness of knowledge is found in a dynamic equilibrium process; the drawing of the line between what proves to be knowledge and what does not prove to be knowledge is therefore dynamic as well; (c) knowledge is now described as inherently dynamic: it is a snapshot of permanently evolving processes; it changes, it can be revised, it can become dominant, it can become outdated or it can be forgotten.[11]

Apparently, the constructional focus manages to solve many of the problems of the traditional, reconstructional notion of knowledge. As a dynamic concept, the revised notion of knowledge is able to depict evolution and change. By considering cognitive processes in general, other symbol processes come into focus. Knowledge is no longer limited to sentences and language systems but embraces all symbol systems. These appear in many areas and they occur in many media. Therefore, it is a misunderstanding to limit the notion of knowledge to a fixed branch of science. It rather relates to certain structures and systematizations. Finally, the concept of rightness releases the debate on knowledge from exaggerated demands of an idealized certainty and absoluteness. The focus on the actual usage and processes of production leads to a more pragmatic point of view.

---

[10]Goodman and Elgin propose another usage of terminology, which does not prove suitable for the purposes here. They apply the notion of understanding in three different ways: as a skill, as a process, and as "what the cognitive process achieves" (Goodman and Elgin, 1988, pp. 161 f.).

[11]A certain type of dynamics could also be attributed to the traditional theory of knowledge, but it differs significantly from the dynamic concept introduced by Goodman and Elgin. The former is the dynamic of growth as an accumulation of knowledge within a static system. The latter sets the system itself in motion, the system itself is under permanent reconception. Within this model, it becomes possible to explain changes in ordering, weighing or focus, modification of criteria of rightness and so on.

What is of major interest here are the consequences for the dogmatic distinction into context of justification and context of discovery. The lesson we learn from Goodman's critique and his model of justification is far-reaching: in short, it says that discovery matters for justification. The process of gaining insights turns out to be the justification of those insights at the same time. Searching, probing, scrutinizing, testing, and assessing go hand in hand. Hence, we have the astonishing result that in order to solve problems related to the context of justification we need to consider the context of discovery. If we take the findings of Goodman and Elgin seriously, the separation of the context distinction turns out to be a severe misunderstanding of the epistemic analysis of knowing; as genesis shows up as a part of justification.

Nevertheless, Goodman's approach not only helps to identiy how the context distinction leads us on the wrong track but also gives us a far better understanding of the method of reconstruction. In contrast to ways of constructing, reconstruction focuses on the results of the epistemic genesis, which are brought in an idealized, abstracted and formalized system. Reconstruction works with the results of the construction process. It simplifies and emphasizes, it draws sharp boundaries and distinctions in order to attain a better understanding. Many manifestations of knowledge can serve as an example: encyclopedias, textbooks, symbolic or pictorial representations. As shown above, there are important interrelations between methods of reconstruction and processes of construction. If one considers the results in isolation, severe misunderstandings can result. Only if we investigate them within their context of construction do they lead to a consistent theory of knowledge.

5 An outlook: The reconciliation of philosophy and sociology of science within a revised theory of knowledge

Given the complex relation between ways of constructing and the method of reconstruction that leads to a reconciliation of the context of discovery and the context of justification, what are the consequences for philosophy and sociology of science? In Goodman's writings, we learn much about how to gain a better understanding and how to justify these findings, but his analysis remains limited to aspects of symbolic functioning and characteristics of symbol systems. However, in order to discuss the consequences for the relation of philosophy and sociology of science, we need to know more about the impact of historical and cultural influences on dynamic epistemic processes. Therefore, the challenge for further investigations is to analyse in detail the epistemic genesis, to question which circumstances become influential and which aspects intervene epistemically. Already a first glimpse tells us that creative inventions and novelties play a crucial role in epistemic genesis. Adjustment, modification, improvement, testing, performing, and so on seem to be individually set off. Is epistemic genesis in the end driven by individual novelties?

For the theory of knowledge, these consequences turn out to be a major challenge heralding substantial change. At the core of the theory of knowledge, individual processes become visible. Nevertheless, it is crucial to notice that the focus on individual, dynamic and active processes is not equivalent to an epistemic solipsism. Individually triggered mutations and dynamics do not happen in isolation. Those processes are inseparable from a transindividual public sphere. There is a permanent interrelation between the individual and its life-world, which has an effect in both directions. By this, we gain on the one hand latitude for negotiation, for exchange and influence on a personal level. On the other hand, we can explain rules, practices, and usage beyond an individualized validity. We therefore have to shift the perspective. Starting points are singular processes of construction within a dynamic relationship with other individuals, in context, practice and usage. Those settings become part of the epistemic genesis and become influential for the results of the process. By this, cultural, historical and sociological aspects enter into theory of knowledge. In turn, knowledge can be investigated in its cultural, historical and social embedding.

Therefore, these readjustments not only help to overcome persistent obstacles in epistemology but they also open novel perspectives in research. To give an example: on this ground, it becomes possible to develop the epistemological framework for investigating questions concerning the *dynamics of epistemic diversity*. To this day, we have difficulties to explain how different forms of knowledge relate to each other.[12] In the succession of pluralistic approaches we have some means at hand to characterize specific kinds of knowledge. However, based on a revised theory of knowledge, it will be possible to describe the intersections and interactions between different kinds of knowledge, which remains a desideratum so far. When local knowledge is confronted with scientific knowledge, when expert knowledge meets the knowledge of a layman, when technical knowledge gets involved with everyday knowledge, they can trigger processes of mutual change, marginalisation, emancipation, alteration or reinforcement. In order to describe these kinds of processes from an epistemic point of view, a theory of knowledge is needed that is grounded both in a pluralistic and dynamic concept of

---

[12] E.g., Weingart (2003, p. 141) points to the necessity of exploring the "inferences" between scientific and other forms of knowledge; Böschen and Schulz-Schaeffer (2003, pp. 210 f.) and Böschen and Wehling (2004, p. 20) focus on the "areas of interaction, overlapping, and conflict" of different forms of knowledge as a novel research agenda; Maasen (2009, p. 81) mentions the "interaction of different forms of knowledge" as a challenge for theory.

knowledge, and which is able to capture historicity and contingency as well as societal influences on knowledge (cf. Ammon, 2007; 2009, especially pp. 178 ff.).

These insights have an important impact on the relation of philosophy and sociology of science. When philosophy of science starts to acknowledge that the context of discovery matters epistemically, we have built the foundations for tackling the gap. Within the setting of traditional theory of knowledge, it was standard practice to investigate philosophical theory of knowledge in a decontextualized manner on an abstract level of formalized systems. With the processes of genesis as the new perspective in epistemology, related topics such as creativity, actions and interests, social, political and historical conditions come into focus as well. However, this is not equivalent to the thesis that knowledge is a purely social construction. It only leads to a reassessment of individual and social factors in theory of knowledge. By this, looking at reconstruction and construction reassesses the relationship of philosophy and sociology of science. Within the new setting, they have found common ground. If this leads to an exchange of concepts and theories, it can be the basis for bridging the gap in the near future.

Bibliography

Ammon, S. (2007). Wissensverhältnisse im Fokus. Eine erkenntnistheoretische Skizze zum Post-Pluralismus. In Ammon et al. (2007b), pages 59–77.

Ammon, S. (2009). *Wissen verstehen. Perspektiven einer prozessualen Theorie der Erkenntnis.* Velbrück Wissenschaft, Weilerswist.

Ammon, S., Heineke, C., and Selbmann, K. (2007a). Einleitung. In Ammon et al. (2007b), pages 9–18.

Ammon, S., Heineke, C., and Selbmann, K., editors (2007b). *Wissen in Bewegung. Vielfalt und Hegemonie in der Wissensgesellschaft.* Velbrück Wissenschaft, Weilerswist.

Baumann, P. (2002). *Erkenntnistheorie. Lehrbuch Philosophie.* Metzler, Stuttgart.

Böschen, S. and Schulz-Schaeffer, I., editors (2003). *Wissenschaft in der Wissensgesellschaft.* Westdeutscher Verlag, Wiesbaden.

Böschen, S. and Wehling, P. (2004). Einleitung: Wissenschaft am Beginn des 21. Jahrhunderts. Neue Herausforderungen für Wissenschaftsforschung und -politik. In Böschen, S. and Wehling, P., editors, *Wis-*

senschaft zwischen Folgenverantwortung und Nichtwissen. Aktuelle Perspektiven der Wissenschaftsforschung, pages 9–33, Wiesbaden. Verlag für Sozialwissenschaften.

Carnap, R. (1928). *Der logische Aufbau der Welt*. Weltkreis-Verlag, Berlin-Schlachtensee. Page numbers refer to the English translation (Carnap, 2003).

Carnap, R. (2003). *The logical structure of the world and pseudoproblems in philosophy*. Open Court, Chicago and La Salle, IL. Translated by Rolf A. George.

Elgin, C. Z. (1996). *Considered judgment*. Princeton University Press, Princeton, NJ.

Elgin, C. Z. (1999). Epistemology's end, pedagogy's prospects. *Facta Philosophica*, 1(1):39–54.

Franklin, A. (1986). *The neglect of experiment*. Cambridge University Press, Cambridge.

Fuller, S. (1988). *Social epistemology*. Indiana University Press, Bloomington, IN.

Goldman, A. I. (1999). *Knowledge in a social world*. Oxford University Press, Oxford.

Goodman, N. (1951). *The structure of appearance*. Harvard University Press, Cambridge, MA.

Goodman, N. (1954). *Fact, fiction, and forecast*. Athlone, London.

Goodman, N. (1978). *Ways of worldmaking*. Hackett, Indianapolis, IN.

Goodman, N. and Elgin, C. Z. (1988). *Reconceptions in philosophy and other arts and sciences*. Hackett, Indianapolis, IN.

Gottschalk-Mazouz, N. (2007). Was ist Wissen? Überlegungen zu einem Komplexbegriff an der Schnittstelle von Philosophie und Sozialwissenschaften. In Ammon et al. (2007b), pages 21–40.

Hacking, I. (1983). *Representing and intervening. Introductory topics in the philosophy of natural science*. Cambridge University Press, Cambridge.

Howard, D. (2006). Lost wanderers in the forest of knowledge: Some thoughts on the discovery–justification distinction. In Schickore and Steinle (2006b), pages 3–22.

Hoyningen-Huene, P. (1987). Context of discovery and context of justification. *Studies in History and Philosophy of Science*, 18(4):501–515.

Hoyningen-Huene, P. (2006). Context of discovery versus context of justification and Thomas Kuhn. In Schickore and Steinle (2006b), pages 119–131.

Knoblauch, H. (2005). *Wissenssoziologie*. UVK, Konstanz.

Longino, H. E. (2002). *The fate of knowledge*. Princeton University Press, Princeton, NJ.

Maasen, S. (2009). *Wissenssoziologie*. transcript Verlag, Bielefeld, 2nd edition.

Nickles, T. (1980a). Introductory essay: Scientific discovery and the future philosophy of science. In Nickles, T., editor, *Scientific discovery, logic, and rationality*, pages 1–59, Dordrecht. Reidel.

Nickles, T. (1980b). *Scientific discovery: Case studies*. Reidel, Dordrecht.

Rawls, J. (1971). *A theory of justice*. Belknap Press of Harvard University Press, Cambridge, MA.

Reichenbach, H. (1966). *Experience and prediction. An analysis of the foundations and the structure of knowledge*. University of Chicago Press, Chicago, IL, 6th edition.

Rheinberger, H.-J. (1997). *Toward a history of epistemic things. Synthesizing proteins in the test tube*. Stanford University Press, Stanford, CA.

Richardson, A. (2006). Freedom in a scientific society: Reading the context of Reichenbach's contexts. In Schickore and Steinle (2006b), pages 41–54.

Schickore, J. and Steinle, F. (2006a). Introduction: Revisiting the context distinction. In Schickore and Steinle (2006b), pages vii–xix.

Schickore, J. and Steinle, F., editors (2006b). *Revisiting discovery and justification. Historical and philosophical perspectives on the context distinction*, volume 14 of *Archimedes*. Springer, Dordrecht.

Schiemann, G. (2006). Inductive justification and discovery. On Hans Reichenbach's foundation of the autonomy of the philosophy of science. In Schickore and Steinle (2006b), pages 23–39.

Sintonen, M. and Kiikeri, M. (2004). Scientific discovery. In Niiniluto, I., Sintonen, M., and Wolenski, J., editors, *Handbook of epistemology*, pages 205–253, Dordrecht. Kluwer.

Steup, M. (2008). The analysis of knowledge. In Zalta, E. N., editor, *The Stanford encyclopedia of philosophy. Fall 2008 edition.*

Weingart, P. (2003). *Wissenschaftssoziologie.* transcript Verlag, Bielefeld.

François, K., Löwe, B., Müller, T., Van Kerkhove, B., editors,
Foundations of the Formal Sciences VII
Bringing together Philosophy and Sociology of Science

Science as socially distributed cognition: Bridging philosophy and sociology of science

MATTHEW J. BROWN[*]

School of Arts & Humanities, The University of Texas at Dallas, 800 W Campbell Road, JO31, Richardson, TX 75080, United States of America.
E-mail: mattbrown@utdallas.edu

In this paper, I want to make the following claim plausible:

> Analyzing scientific inquiry as a species of socially distributed cognition has a variety of advantages for science studies, among them the prospects of bringing together philosophy and sociology of science.

This is not a particularly novel claim; indeed, Paul Thagard has been suggesting something like this for well over a decade, while philosophers like Ronald Giere and Barton Moffat have been stumping for the distributed cognition approach in more recent years, and Nancy Nersessian's Cognition and Learning in Interdisciplinary Cultures research group at the Georgia Institute of Technology has been fruitfully applying this approach to the study of research laboratories and other scientific institutions.

I will retrace some of the major steps that have been made in the pursuit of a distributed cognition approach to science studies, paying special attention to the promise that such an approach holds out for bridging the rift between philosophy and the social studies of science. Such an approach is not without its pitfalls, and I will consider several problems, both for distributed cognition as a theory and for its applications to science. I will argue that there is a path out of the woods, and try to point the way. Ultimately, I argue that we shall have to widen the scope of the distributed-cognition approach.

---

[*]Many thanks to Michael Cole, Edwin Hutchins, Nancy Nersessian, and Paul Churchland for encouraging and discussing this work. Thanks also to P. D. Magnus, for helpful discussions about d-cog and science, to the members of the UCSD D-Cog/HCI Lab for allowing me to present an earlier version of this project, and to the organizers of the *Foundations of the Formal Sciences VII* conference, especially Benedikt Löwe, who encouraged and helped me to make the long journey from San Diego to Brussels.

Received by the editors: 30 January 2009.
Accepted for publication: 20 August 2009.

1 What is d-cog?

Distributed cognition (d-cog) is a radical theory in cognitive science, created primarily by researchers at the University of California, San Diego (UCSD), which maintains that one can fruitfully analyze activities taking place between one or more people along with technological artifacts as *cognitive* in the same way that traditional cognitive science has analyzed certain intrapersonal processes. The beginnings of the approach can be seen in the *Parallel Distributed Processing* research group on connectionist (neural-network) models of cognition. When it came to an explanation of how a neural-network architecture can do science, mathematics, and logic, they made an intriguing suggestion:

> If the human information-processing system carries out its computations by "settling" into a solution rather than applying logical operations, why are humans so intelligent? How can we do science, mathematics, logic, etc.? How can we do logic if our basic operations are not logical at all? We suspect the answer comes from our ability to create artifacts—that is, our ability to create physical representations that we can manipulate in simple ways to get answers to very difficult and abstract problems. (Rumelhart et al., 1987, p. 44)

This is quite the break from classical cognitive science research in two ways. First, cognitive science has traditionally treated "cognition" as a matter of computational operations on symbolic states, not unlike the operations of logic or the architecture of an ordinary computer. The move towards a connectionist architecture, where the basic computational processes are more like pattern-recognition than applying logical rules, is the radical step that the *Parallel Distributed Processing* group was most keen to argue for. Second, the cognitive sciences ordinarily focus on what goes on with an individual person, and "cognition" is what goes on in their head. It is this second break that spurs d-cog.

D-cog takes a wider perspective than classical cognitive science. It is concerned with use of "cognitive artifacts" such as pen and paper, longhand mathematical calculation, and digital computers. It is also interested in the cognitive role of social interactions, cultural institutions, and forms of social organization. D-cog attempts to move the boundaries of our concept of "cognitive activity" out from the head and into the world of social and technological interactions. The foundational text for research in distributed cognition is *Cognition in the Wild,* the work of another UCSD cognitive scientist, Edwin Hutchins, who was trained in cognitive anthropology (Hutchins, 1995a).

There are two ways we might understand the project of distributed cognition research. A more cautious definition—the one preferred by Giere, for example[1]—would be that some socially and technologically distributed activities can profitably be understood as "cognitive", while allowing that many elements of cognition—including agency—remain in the realm of the individual. The more radical definition of d-cog that one could adopt, and the one most supported by Hutchins's own statements, is more sweeping; as Hutchins says, "I hope to show that human cognition [...] is *in a very fundamental sense* a cultural and social process" (1995a, p. xiv, emphasis mine).

Two examples have become pervasive in papers about d-cog.[2] Nevertheless, it is worth briefly discussing each. The first originates in (Rumelhart et al., 1987). When we multiply large numbers, we rarely if ever do it in our heads. With a pencil and paper, multiplying even very large numbers is transformed into a simple task, requiring no more than the ability to do one-digit multiplication and addition. A nearly impossible task for an individual human cognitive system becomes perfectly easy when distributed across the human-pencil-paper system (cf. Figure 1).

```
        12044
   x      432
   _____
        24088
        36132
        48176
   _____
      5203008
```

FIGURE 1. Longhand Multiplication

A second key example comes from Hutchins's work on ship navigation in the US Navy (1995a). Navigation on a large naval vessel is not the job of a single individual (as it is for Hutchins's contrast case of the Micronesian canoer), but rather the work of a team of people performing various jobs using various instruments. Here is a somewhat simplified account: A crewman (called a pelorus operator) is given a landmark to identify by a plotter in the pilothouse. The pelorus operator then uses a piece of equipment called an "alidade" to determine the bearing of the landmark. Generally, there

---

[1] Giere (2002a, p. 6); Giere (2002b, p. 238); Giere (2002c, p. 295); Giere (2006, pp. 112–114); Giere and Moffatt (2003, p. 4).

[2] Cf. Hutchins (1995a); Giere and Moffatt (2003); Magnus (2007).

is more than one pelorus operator, and they all relay their information to a plotter. The plotter uses that information to determine the location of the ship and its bearing. The plotter relies on a specially structured map, compasses and protractors, etc. in order to use the information about the bearing of landmarks to compute the ship's location and bearing.

In this example, it is physically impossible for a single human being (given the size of the ship, the location of various vantage points, and the time in which the task must be completed) to do the cognitive work of figuring out the location and bearing of the vessel. Of course, on a *different* kind of ship, it is possible for a single person, and indeed, in the case of the Micronesian navigator, it is possible for a lone individual to do so without instrumentation. Nevertheless, the navigation team on a large naval vessel completes the cognitive task as a team using artifacts. The essence of a d-cog analysis is in treating this network of individuals and artifacts as a single cognitive system.

2 Science as d-cog

Can science be analyzed using d-cog? Consider a case discussed by Giere and Moffatt (2003), originally due to Dumas (1834) as discussed by Klein (1999). Chemical formulae were originally introduced by Jacob Berzelius in 1813 (Klein, 2001, p. 7). A Berzelian formula like the one in Figure 2 allows one to do theoretical chemistry by manipulating such symbols on paper, replacing the need to directly manipulate chemicals. All one needs to do in order to determine what is going on in a reaction, knowing something about the products and the reactants, is assume conservation and balance the equation. Just as doing long multiplication by hand transforms a complex calculation into a set of simple pattern-matching problems, so the use of chemical formulae as a cognitive artifact transforms the complex theoretical or experimental analysis into a simple exercise in pattern matching (cf. Figure 2).

$$C_8H_8 + H_4O_2 + Ch_4 = C_8H_8O_2 + Ch_4H_4 \ [sic]$$

FIGURE 2. Chemical Formula for reaction of alcohol and chlorine according to Dumas (1834)

Consider another example, the Hubble Space Telescope, an important piece of scientific equipment in contemporary research in astronomy, astrophysics, and cosmology (Giere, 2006, p. 99–100). The telescope is a large and complex instrument that must be operated remotely. It is used by an organized group of people, and that use is mediated by further instruments and computer equipment on earth. To draw out the sense in which d-cog analysis is appropriate, think of the telescope as the eyes of a large cognitive system that also includes the group of scientists and the earth-bound computer equipment. Just as cognitive science can study ordinary perception, distributed cognitive science can look at this distributed system of "perception."

A final example, due to Nancy Nersessian and her collaborators at Georgia Tech (Nersessian et al., 2004) comes from the cognitive-ethnographic study of research in a biomedical engineering laboratory. Nersessian discusses how certain lab equipment are used to *model* actual biological processes. For example, the lab she studies uses a piece of equipment called a "bioreactor," which, among other things, models blood flow in a way which can be used to study arterial cells and biomedical devices. Nersessian's work explicitly treats the bioreactor, along with the skills one needs in order to use it in certain ways as a "mental model" for the distributed system of the biomedical laboratory. Doing so reveals interesting facts about the system that aren't available if you treat it just like a device or an instrument.[3]

3 The cognitive and the social

Latour and Woolgar (1986) issued their famous "ten-year moratorium on cognitive explanations of science," promising "that if anything remains to be explained at the end of this period, we too will turn to the mind!" (Latour and Woolgar, 1986, p. 280; quoted by Giere and Moffatt, 2003, p. 301). Of course, the moratorium has run out, much remains to be explained, and they never turned solely to the mind to provide the missing explanations—but that's not the point. What is interesting are the motivations and implicit assumptions behind this rhetorical flourish.

Part of the reason they issued such a moratorium, as Giere and Moffatt (2003, p. 301), Nersessian (2005, p. 18), and others have argued, is that they held to a rigid dichotomy of cognitive and social factors. Because their primary goal was to get a serious sociology of science going, they regarded such a moratorium as necessary. In order to make room for social explanations of science, we must, they thought, bracket all cognitive issues and explanations.

D-cog provides an alternative to this way of thinking. It shows us how to treat the cognitive and the social as the same thing for certain purposes. Because cognitive structures need not exist only in the mind (and perhaps never do so, if the radical version of d-cog is correct), but instead can exist in the complex interactions of social groups and technological artifacts, one can study social groups cognitively, or cognitive systems sociologically. There need be no unbridgeable divide between social and cognitive explanations.[4]

---

[3]I will return to this case in further detail below, to indicate some of the major gains of such an analysis.
[4]If I read him correctly, Bruno Latour has come around to this more sophisticated

What's most interesting about the possibility of seeing the social in terms of the cognitive and vice versa is that it might just help heal the wounds of the Science Wars and bring the various parts of science studies which are often at loggerheads—especially philosophy and sociology of science—together towards a more common purpose. Because of the perceived incompatibility of the cognitive and the social, the terms of analysis of much recent sociology of science—negotiation, authority, power, mobilizing resources—seem to have a cynical cast, dismissive of the virtues of science. By contrast, the normative concerns of philosophers of science—justification, realism, objectivity—seem divorced from the obvious social reality of science.

There are plenty of philosophers nowadays—such as Helen Longino (2002a; 2002b; 2002c) and Philip Kitcher (2001, 2002a,b)—trying to reconcile the cognitive and the social, the normative issues of philosophy of science with descriptive sociological analyses. Their arguments are mainly about the very possibility of such a reconciliation, and focus more on the reformulation of traditional philosophical issues (e.g., objectivity) in ways that involve social relations and institutions rather than focusing on the properties of individual scientists or the *abstract structure* of science. D-cog presents more than the mere possibility of an in-principle or a post-hoc reconciliation. It allows one to fruitfully re-interpret the excellent and extensive body of sociological and historical studies in line with cognitive and epistemic concerns.

Consider again the case of chemical formulae. Bruno Latour (1986) has emphasized the importance of such innovations in the history of science. According to Giere and Moffatt, Latour thinks that something like chemical formulae are important because they concentrate information in a way that

> confers authority and power on those who control it. And it leads others to align themselves with such powers, thus increasing still further their authority and power. In a struggle for dominance, whether in science, politics, or war, those with the most and strongest allies win. (Giere and Moffatt, 2003, p. 305)[5]

What d-cog allows Giere and Moffatt to do is to look at the specifics of Latour's analysis of the social-technological aspects of science and point out the *cognitive* function of various parts of the process. What might be cast

---

view of the cognitive-social relation. Cf. Latour (1991, 1996).

[5] Having read Latour, this seems a slight exaggeration of his point. Already he seems to recognize the d-cog perspective when he says: "An average mind or an average man, with the same perceptual abilities, within normal social conditions, will generate totally different output depending on whether his or her average skills apply to the confusing world or to inscriptions" (Latour, 1986, p. 22). I take this to imply that inscriptions function as a cognitive artifact that change the functioning of the mind independent of the particular agent's perceptual abilities.

by a sociologist in terms of exerting power and gaining allies can be cast in terms of improving cognitive capacities of a distributed system over a naked cognitive agent. As Giere and Moffatt say about this particular case,

> The invention of new forms of external representation and of new instruments for producing various kinds of representations has played, and continues to play, a large role in the development of the sciences. From a cognitive science perspective, both sorts of invention amount to the creation of new systems of distributed cognitive system. So, for us, the notion of distributed cognition brings under one category such things as Cartesian coordinates and the telescope, both of which are widely cited as major contributions to the Scientific Revolution. (Giere and Moffatt, 2003, p. 305)

Even more promising than the idea of reinterpreting sociological and historical work in cognitive-epistemic terms is empirical work being done by cognitive scientists, philosophers of science, and researchers in science studies *using* the methods and theoretical frameworks of d-cog to analyze science. To return to another example from above, Nersessian and her collaborators have been studying work in a biomedical engineering laboratory, applying cognitive, historical, and ethnographic methods and understanding the organization of the lab and the function of artifacts within the lab as parts of a distributive cognitive system. Such an analysis allows researchers to understand how a bioreactor is both a "significant cultural artifact... [and] a locus for social interaction" (Nersessian, 2005, p. 50) with a history of different kinds of roles in the culture of the laboratory and also as a model that plays a role in distinctive types of representation and reasoning. Without such an analysis, the fact that a single object plays both of these roles (and the pervasiveness of such objects in science) is a colossal coincidence and a total mystery. One might even be lead to deny that the object has an important cognitive side (as sociologists of science are often led to do) or to claim that its cultural history and social roles are inessential to its role in representation and reasoning (as philosophers have often done). D-cog analysis makes better sense of what is going on in such cases, and makes better sense of how the social and the cognitive are integrating in science as a whole.

4 Challenges

Applying Hutchins's d-cog theory to the study of science is not without its problems. I will focus on two major challenges to the applicability of the theory.

The first problem is that d-cog looks like a theory applicable to fairly static systems.[6] The paradigm applications of d-cog in Hutchins's work

---

[6] I believe this problem was first pointed out to me by Yrjö Engeström in conversation.

(1995a; 1995b)—airplane cockpits determining their speed, crews of Navy ships navigating through a harbor, even pencil-and-paper multiplication—respond to dynamic situations where "problem-solving situations change in time" (Nersessian, 2005, p. 36), but the organization and nature of the technological artifacts in play are treated as static. The analysis is thus "dynamic but largely *synchronic*" (Nersessian, 2005, p. 36). But really, these systems are evolving, if slowly, both from external pressures (invention of new technology, new safety protocols, changes in policy) and internal developments (shifts in Navy culture, new pilots gaining skills, invention of new techniques). Further, many other kinds of cognitive activities are much more *diachronically* dynamic, involving creativity, innovation, and rapid changes in technology and social structure.[7] This is especially true of systems like scientific laboratories, where innovation, new discovery, and creative problem-solving are essential parts of the activity. Another aspect of this problem is that d-cog analyses tend to treat relatively well-bounded systems, with low-bandwidth information flow from outside the system and high-bandwidth information flow within the system. In order to straightforwardly apply Hutchins's d-cog framework, the nature of the task at hand and the system that carries it out must be rather well-bounded. On the other hand, many activities, including scientific activities, have quite vague and porous boundaries. What counts as part of the system might change rapidly as the activity goes on.

The second problem comes from a direct critique of Giere's appeals for treating science as d-cog by Magnus (2007). Magnus's critique turns on a particular move in Hutchins's (1995a) account of d-cog, where he relies on the tripartite distinction from Marr (1982) between *computation*, *algorithm*, and *implementation* (Hutchins, 1995a, pp. 50–52), which Magnus simplifies into the distinction between *task* and *process*,[8] where *task* is an abstract description of the computational goal or behavior that the cognitive system is to satisfy, and the *process* is just a specification of how the task is to be accomplished. This furnishes Magnus with a compellingly succinct definition of d-cog:

> An activity counts as d-cog only if the process is not enclosed by the epidermis of the people involved in carrying out the task. The implementation uses tools and social structures to do some of the cognitive work. (Magnus, 2007, p. 299)

Where the task in question "would be cognitive if the process were contained entirely within the epidermis of one individual" (Magnus, 2007, p. 300) .

---

[7]"Although there are loci of stability, during problem-solving processes the components of the systems undergo development and change over time" (Nersessian, 2005, p. 36).

[8]He is following Ron McClamrock (1991); cf. also McClamrock (1995, § 1.3).

So, an activity is d-cog only if the process is not located inside the skin of an individual carrying out a cognitive task. Is science like that? It is easy enough to see that on Magnus's interpretation, if we are going to be able to analyze scientific activity as a species of d-cog, we must be able not only to analyze the scientific *process*, but we must also be able to specify the *task* of science. This poses two types of concern. First, at a local level, can we always abstractly specify a *task* for science? Does a biomedical engineering laboratory have a well-specified task? Does a physics journal? What about a conference on global warming? While it seems likely that there are some scientific activities which might be amenable to such an analysis, it seems dubious that one could specify the kind of *computational* task necessary for d-cog analysis for all or even most scientific activities.

The second worry that Magnus raises is whether one can specify a *global* task for science, and thus do a d-cog analysis of science "writ large". That is, "Can we understand science altogether as one giant, distributed cognitive enterprise?" Such an interpretation is already suggested by Hutchins (1991, p. 288). It would be a lucky thing if we can do so, for we could then give a clear explanation of the common view in science studies that it is the large-scale institution of science, rather than individual scientists, which produce or are responsible for scientific knowledge. To this end, Magnus analyzes three candidates for giving a *task* analysis for science-as-a-totality: Merton's *ethos* of science, Philip Kitcher's ideal of the distribution of cognitive labor, and his more recent image of well-ordered science.

As you might imagine, the prognosis is dire; Magnus is rightfully pessimistic about the possibility of specifying the task of science-as-such. After all, the range of activities of science, the differences in approaches in different research traditions, the variety of uses to which science is put, and so on make it highly unlikely that there is one simple task that all of science aims at. One need only compare high-energy physics to molecular biology to pharmaceutical trials to see how unlikely such a project is.[9] Given the poor prospects of rescuing d-cog analyses of science in this way, I will suggest we look elsewhere.

5 Prospects for a d-cog theory of science

Here's where we stand: d-cog holds great promise for analyzing science in a way that makes the relation of the social-technological nature of science to its cognitive-epistemic virtues most perspicuous, and thus joining together what the Science Wars hath put asunder, of healing the rift between philosophy and sociology of science. However, using d-cog to analyze science faces some severe difficulties: it treats systems whose basic structures and

---

[9]Cf. Knorr-Cetina (1999); Giere (2002a).

resources are fairly static, while science is not only synchronically but diachronically dynamic. Scientific systems evolve, and d-cog provides little in the way of resources for analyzing that evolution. D-cog applies to well-bounded systems, whereas the boundaries of science aren't so clear. D-cog analysis requires the specification of a computational task that can be implemented by a distributed process, while it seems doubtful that a global task can be specified for science, and even unlikely that a more local task can be specified for many important cases. Does this spell doom for the d-cog approach to science studies? Is there any way forward?

I think there is, and that way depends most importantly on going well beyond Hutchins's work from the mid-1990s. Of course, Hutchins himself is an active researcher and has gone beyond that work himself, into areas like conceptual change, learning, and the embodiment of cognition. Likewise, Nersessian, for example, relies on d-cog, but has taken it beyond Hutchins's original formulations. There are also traditions and research programs related to d-cog, such as neo-Vygotskian psychology, cultural-historical activity theory, and situated action theory, that have things to offer a *broadly* d-cog account of science. On the basis of the criticisms so far discussed, I will conclude by indicating the ways we must modify our understanding of d-cog in order for it to have positive prospects as an account of science.

The first important point to make, as against Magnus's interpretation and some of Hutchins's formulations of d-cog, is that cognition is *not* computation (in the classical sense). Certainly, computation is one kind of thing that cognitive agents and cognitive systems do, but it isn't the case that cognition is identical to computation.[10] Cognition is not a single algorithm or program, though it may use algorithms. Human cognitive capacities at their best are flexible and responsive to particular situations, creative and dynamic. Cognition is a multi-purpose capacity in humans, and likewise in any other sort of cognitive system.[11] While this may be a controversial point in some circles in cognitive science, those circles are shrinking precipitously. It is hard today not to agree with the point that was radical when proposed by the *Parallel Distributed Processing* group decades ago, that human cognition bears little if any resemblance to classical computation.[12]

While the approach may be controversial, there may be some valuable lessons to be learned from cultural-historical activity theory (CHAT) for

---

[10] Alternatively, as an anonymous referee pointed out to me, one might regard this as a radical extension of the idea of *computation*, rather than a denial of the identity of cognition and computation.

[11] Even those who regard cognition as having a modular architecture must admit that the human cognitive system at large is a complicated, multi-purpose, dynamic, and flexible system.

[12] Even if one does not accept *Parallel Distributed Processing*-type models.

providing a d-cog analysis of science.[13] CHAT provides a tripartite distinction between *operations, actions,* and *activities* that adds a useful layer to the talk about task vs. process. Operations are the basic components of actions; they are generally routinized human behaviors or mechanical operations, carried out under certain conditions, instrumental to engaging in some action. Actions are conscious, goal-directed processes, undertaken by individuals or small groups. For example, Leont'ev (1978, p. 66) describes learning to drive a car with a manual transmission. At first, all the processes of driving the car—breaking, using the clutch, shifting gears—require conscious attention. They are the focal, goal-directed activities. For the accomplished driver, these processes become unconscious, subordinated to actions like speeding up, driving up a steep incline, driving to work. In the end, the unconscious operations are actually off-loaded to a machine, the automatic transmission.

Beyond the level of action is the activity. Actions are goal-directed, relatively short-lived and well-bounded in time and space. Activities exist in and evolve over longer periods of time; they have a history. They are associated with a culture or a community, and they are often embedded in institutions or forms of social organization. While actions are simply goal-directed, activities are aimed at a more general, less-bounded, and changeable object or motive. While the particular actions of a welder in a factory have a quite well-defined goal (joining two metal pieces together), the activity of the whole factory has a more nebulous object of gaining profit, and the way that motive is conceived over time may change (for example, a change to a more socially-conscious, "green" corporate mission may alter both the ways that profit is got and the way that gaining profit is understood). The object of the activity system need not be at all available to the individual members of the system; indeed, the workers need not have any ideas about the economic purposes of the factory—they need only be in it to get a paycheck for themselves.

This set of distinctions may prove fruitful for thinking about science as d-cog. In particular, the *task* as Magnus (2007) seems to understand it seems identical to the *goal* to which actions are directed. The task-process distinction may thus make perfect sense at the level of action, but to get the whole sense in which science is a d-cog *activity*, we may need to think of it at the level of activity directed at an object which is partially constituted by the evolution of the activity itself.

Another potentially necessary turn is to supplement Hutchins's cognitive ethnography with Nersessian's cognitive-historical method. Nersessian is keenly aware of the problem of evolving systems for Hutchins's (1995a;

---

[13]Cf. Leont'ev (1978); Engeström (1987); Cole and Engeström (1993); Cole (1988); Engeström et al. (1999).

1995b) account, especially as applied to science. Indeed, the argument that Hutchins's account does not naturally accommodate the evolution of cognitive systems I gave above is her argument. In their own d-cog research on biomedical labs, Nersessian and her collaborators combine ethnographic investigation of the particular system with *cognitive-historical* analysis, which looks at different scales of history to understand the evolution of problems, concepts, cognitive artifacts, etc. (Nersessian et al., 2004).

So, *is* science a distributed cognitive system? This has been challenged on the basis of it being an evolving, messy, less-bounded system. Magnus (2007) has challenged it on the basis of whether there is a particular task that science carries out. But *what is a cognitive system* anyhow, even in the traditional sense of "cognitive system?" This shouldn't stand or fall on the details of a certain framework of cognitive analysis. After all, presumably, *I* am some kind of cognitive system, even though I am not built to carry out one specific and well-bounded task, even though my cognitive activities evolve, and aren't always as well-bounded as certain cognitive theories might presuppose. Certainly, the limitations of a particular approach to d-cog shouldn't disqualify the more general notion. Rather, this points the way towards the need for better, more complex models of distributed cognition that might do a better job of applying to science. I have gestured towards some possibilities that seem particularly fruitful in the face of these difficulties. There is much more work to be done, and the possibilities are inspiring.

Bibliography

Cole, M. (1988). Cross-cultural research in the sociohistorical tradition. *Human Development*, 31:137–151.

Cole, M. and Engeström, Y. (1993). A cultural-historical interpretation of distributed cognition. In Salomon, G., editor, *Distributed cognitions. Psychological and educational perspectives*, Learning in doing. Social, cognitive, and computational perspectives, pages 1–46, Cambridge. Cambridge University Press.

Dumas, J. B. (1834). Recherche de chimie organique. *Annales de Chimie de Physique*, 56:113–150.

Engeström, Y. (1987). *Learning by expanding: An activity-theoretical approach to developmental research*. Orienta-Konsultit, Helsinki.

Engeström, Y., Miettinen, R., and Punamäki, R.-L. (1999). *Perspectives on activity theory*. Learning in doing. Social, cognitive, and computational perspectives. Cambridge University Press, Cambridge.

Giere, R. (2002a). Distributed cognition in epistemic cultures. *Philosophy of Science*, 69:637–644.

Giere, R. (2002b). Models as parts of distributed cognitive systems. In Magnani, L. and Nersessian, N., editors, *Model-based reasoning: Science, technology, values*, pages 227–241, Dordrecht. Kluwer.

Giere, R. (2002c). Scientific cognition as distributed cognition. In Carruthers, P., Stitch, S., and Siegal, M., editors, *Cognitive bases of science*, pages 285–299, Cambridge. Cambridge University Press.

Giere, R. (2006). *Scientific perspectivism*. University of Chicago Press, Chicago, IL.

Giere, R. and Moffatt, B. (2003). Distributed cognition: Where the cognitive and the social merge. *Social Studies of Science*, 33:301–310.

Hutchins, E. (1991). The social organization of distributed cognition. Perspectives on socially shared cognition. In Resnick, L. B., Levine, J. M., and Teasley, S. D., editors, *Perspectives on socially shared cognition*, volume XIII of *APA Science Volume Series*, pages 283–307, Washington, DC. American Psychological Association.

Hutchins, E. (1995a). *Cognition in the wild*. MIT Press, Cambridge, MA.

Hutchins, E. (1995b). How a cockpit remembers its speeds. *Cognitive Science*, 19:265–288.

Kitcher, P. (2001). *Science, truth, and democracy*. Oxford University Press, Oxford.

Kitcher, P. (2002a). Reply to Helen Longino. *Philosophy of Science*, 69(4):569–572.

Kitcher, P. (2002b). The third way: Reflections in Helen Longino's *The fate of knowledge*. *Philosophy of Science*, 69(4):549–559.

Klein, U. (1999). Techniques of modelling and paper-tools in classical chemistry. In Morgan, M. S. and Morrison, M., editors, *Models as mediators: Perspectives on natural and social science*, pages 146–167, Cambridge. Cambridge University Press.

Klein, U. (2001). Berzelian formulas as paper tools in early-nineteenth century chemistry. *Foundations of Chemisty*, 3(1):7–32.

Knorr-Cetina, K. (1999). *Epistemic cultures: How the sciences make knowledge*. Harvard University Press, Cambridge, MA.

Latour, B. (1986). Visualization and cognition: Thinking with eyes and hands. *Knowledge and Society*, 6:1–40.

Latour, B. (1991). *Nous n'avons jamais été modernes. Essai d'anthropologie symétrique*. La Découverte, Paris. Page numbers refer to the English translation (Latour, 1993).

Latour, B. (1993). *We have never been modern*. Harvard University Press, Cambridge, MA.

Latour, B. (1996). On interobjectivity. *Mind, Culture, and Activity*, 3(4):228–245.

Latour, B. and Woolgar, S. (1986). *Laboratory life*. Princeton University Press, Princeton, NJ. Second Edition. Introduction by Jonas Salk. With a new postscript by the authors.

Leont'ev, A. N. (1978). *Activity, consciousness, and personality*. Prentice Hall, Englewood Cliffs, NJ.

Longino, H. E. (2002a). *The fate of knowledge*. Princeton University Press, Princeton.

Longino, H. E. (2002b). Reply to Philip Kitcher. *Philosophy of Science*, 69(4):573–577.

Longino, H. E. (2002c). Science and the common good: Thoughts on Philip Kitcher's *Science, truth, and democracy*. *Philosophy of Science*, 69(4):560–568.

Magnus, P. D. (2007). Distributed cognition and the task of science. *Social Studies of Science*, 37(2):297–310.

Marr, D. (1982). *Vision: A computational investigation into the human representation and processing of visual information*. Plenum, New York.

McClamrock, R. (1991). Marr's three levels: A re-evaluation. *Minds and Machines*, 1:185–196.

McClamrock, R. (1995). *Existential Cognition. Computational Minds in the World*. University of Chicago Press, Chicago.

Nersessian, N., Kurz-Milcke, E., Newstetter, W. C., and Davies, J. (2004). Research laboratories as evolving distributed cognitive systems. In Alterman, R. and Kirsch, D., editors, *Proceedings of the 25th annual conference of the Cognitive Science Society. July 31st – August 2nd, 2003. Boston, Massachusetts, USA*, pages 857–862. Taylor & Francis.

Nersessian, N. J. (2005). Interpreting scientific and engineering practices: Integrating the cognitive, social, and cultural dimensions. In Gorman, M. E., Tweney, R. D., Gooding, D. C., and Kincannon, A. P., editors, *Scientific and technological thinking*, pages 17–56, Mahwah NJ. Lawrence Erlbaum.

Rumelhart, D. E., Smolensky, P., McClelland, J. L., and Hinton, G. E. (1987). Schemata and sequential thought processes in PDP models. In Rumelhart, D. E. and McClelland, J. L., editors, *Parallel distributed processing*, volume 2, pages 395–412, Cambridge, MA. MIT Press.

FotFS VII

François, K., Löwe, B., Müller, T., Van Kerkhove, B., editors,
Foundations of the Formal Sciences VII
Bringing together Philosophy and Sociology of Science

Freud's unintended institutional facts

FILIP BUEKENS[1], MAARTEN BOUDRY[2,]*

[1]Tilburg School of Humanities, Tilburg University, P.O. Box 90153, 5000 LE Tilburg, The Netherlands

[2]Vakgroep Wijsbegeerte en Moraalwetenschap, Universiteit Gent, Blandijnberg 2, 9000 Gent, Belgium

E-mail: f.a.i.buekens@uvt.nl; Maarten.Boudry@UGent.be

1 'Applications are always confirmations'

"[A]pplications of analysis are always confirmations of it as well", Sigmund Freud wrote in the *Neue Vorlesungen zur Einführung in die Psychoanalyse*, expressing his confidence that psychoanalysis could be successfully applied as a new autonomous research method in anthropology, literary studies and other disciplines in the humanities (Freud, 1933, *SE XXII*, p. 146), and *de facto* confirmed by countless applications of his theories to cultural and social phenomena.[1] In her insightful study *Freud and the Institution of Psychoanalytic Knowledge*, Sarah Winter points out that Freud's official explanation of the remarkable expansion of the psychoanalytic method was based on two considerations: "According to Freud, the psychoanalytic work in interpreting dreams has shown, in the light of analogies with 'linguistic usage, mythology and folklore', that 'symbols seem to be a fragment of extremely ancient inherited mental equipment' and that 'the use of a common symbolism extends far beyond the use of a common language'" (Winter, 1999, p. 216; referring to *SE VIII*, p. 242). Moreover, Freud stressed the

---

*Numerous people have, in discussions and lectures, commented on this paper. We especially would like to thank Jacques van Rillaer and Frederick Crews, whose work on Freud and the history of psychoanalysis was immensely inspirational for this (and other) papers. The paper summarizes and amplifies an argument developed in a book written by the first author and published in Dutch in 2006. Thanks to Jonathan Biedry for stylistic advice.

[1]All references to Freud's collected works (Strachey, 1959) are indicated in the text by the abbreviation *SE*, followed by the volume (in Roman numerals) and the page (in Arabic numerals).

Received by the editors: 4 March 2009; 21 September 2010.
Accepted for publication: 24 September 2010.

scientific and disciplinary importance of dream research to the field of normal psychology:

> If dreams turned out to be constructed like symptoms, if their explanation required the same assumptions—the repression of impulses, substitutive formations, compromise formation, the dividing of the conscious and the unconscious into various psychical systems—then psychoanalysis was no longer an auxiliary science in the mind of psychopathology, it was rather the starting point of a new and deeper science of the mind which would be equally indispensable for the understanding of the normal. Its postulates and findings could be carried over to other regions of mental happening; a path lay open to it that led far afield, into spheres of universal interest. (*SE XX*, p. 47; also quoted in Winter, 1999, p. 217)

Many theorists still find Freud's 'analytic method'—often referred to by Freud as a 'technique'—useful for interpreting human phenomena. Indeed, almost every anthropological phenomenon has been given numerous and conflicting psychoanalytic interpretations. At the same time, many have sensed that the psychoanalytic method exceeds a threshold, an upper limit beyond which interpretations merely *create* meanings. The psychoanalytic constructions no longer offer revealing insights, they are not supported by independent evidence gathered in other fields of inquiry or by other hermeneutic or scientific methods, and the findings cannot be used to support hypotheses or theories outside the psychoanalytic field proper. When a technique can be used to successfully apply its key theoretical concepts to every human phenomenon, when every version of the theory finds 'confirmations' of its central claims, when none of these confirmations can be productively integrated with findings based on other methods of inquiry, the foundations of the method and its immense success demand a critical explanation. While all hermeneutic theories agree that whatever falls under the label of 'meaning' is a human-created phenomenon, one can still provide a strong argument for the claim that the object of a hermeneutic approach or methodology should not be created by the technique that purports to uncover it. The latter claim cannot simply be dismissed by the trivial fact that every interpretation reflects creative insights of the interpreter.[2] The immense success of psychoanalysis qua hermeneutic method

---

[2] Cf. also Boudry and Buekens (2011). The *lower* threshold is exceeded when intentional concepts and their application methods are reduced to physical or neurological concepts. It is (at least from our perspective) an illusion that mental concepts can be eliminated, but some philosophers have described reductionism as a conceptually coherent possibility. Ironically, Freud himself predicted that his own psychoanalytic findings would eventually be confirmed by neurological research. Cf. Stroud (2004) and Dupré (2004) for accounts of the hidden charm of reductive naturalism. A deeper analysis of

requires an independent explanation that goes beyond Freud's own justifications and, which connects various independently discovered facts about the history of psychoanalysis. In *Le dossier Freud*, a brilliant analysis of the turbulent history of psychoanalysis, Borch-Jacobsen and Shamdasani (2006) show that the self-confirmatory and meaning-productive character of the psychoanalytic technique was a central and persistent objection, which had been levelled against psychoanalysis from its advent. Some of Freud's earliest critics raised this objection, including Alfred Adler, Richard von Krafft-Ebing, Carl Gustav Jung, Albert Moll, and even Freud's one-time friend and intellectual sparring partner, Wilhelm Fliess. Sooner or later, these critics had realized Freud's failure to provide a convincing reply to the objection that psychoanalysis had not legitimated its purported capacity to understand everything. Indeed, the best Freud could offer as a counterargument was the distinctively psychoanalytic gambit that these objections were uniformly motivated by *psychological resistance* to the theory (cf. Borch-Jacobsen and Shamdasani, 2006, Chapter II for an extensive overview). His reply illustrates how apt the objection was: the Freudian concepts and the hermeneutic technique made it possible to even understand its critics and preemptively discount their criticisms. Freud commits himself here to a particularly seductive combination of the *circularity* and *ad hominem* fallacies. Yet despite this criticism levelled at psychoanalysis by psychiatrists, philosophers of science, and historians alike, one intriguing question still remains unanswered: *why* was Freud right when he held that *any* psychoanalytic theory could be used to interpret *any* cultural phenomenon. Why is it that nothing can escape the psychoanalyst's attention? And why does the combination of theory and technique leave, in Frederick Crews' memorable words, "an academic interpreter without even a mathematical chance of having nothing to say" (Crews, 2006, p. 61)?

A meta-analysis of the success of outcomes of the psychoanalytic hermeneutic method should also be able to explain why the interpretations cannot be fruitfully integrated in non-psychoanalytic theories, like neurology (or other evidence-based medicine), sociology, linguistics, and cognitive science. Critics who reject psychoanalysis as a pseudo-hermeneutic (Cioffi, 1998; Macmillan, 1997), or redescribe its therapeutic effects as an outworking of insight placebos (Jopling, 2008), do notice but insufficiently explain its remarkably *closed* character: psychoanalytic theories and interpretations are *impenetrable* by other disciplines (recall Freud's earlier point about psychoanalysis needing 'no independent confirmation') and its interpretations are *irrelevant* in other disciplines. Sociology, cognitive psychology or cultural studies can perfectly neglect psychoanalytic interpretations of phenomena without loss of evidence for their theories.

---

the upper and lower bounds of hermeneutic methods will not be presented in this paper and must await more extensive treatment.

A meta-analysis should also explain why the psychoanalytic method creates the impression of 'having understood' any phenomenon submitted to psychoanalytic treatment. A particularly strong claim in this respect was made by literary theorist Norman Holland, who held that "the phantasy psychoanalysis discovers at the core of a literary work has a special status in our mental life that moral, medieval, or Marxist ideas do not... the crucial point, then, ... is: the psychoanalytic meaning underlies all the others."[3] 'Now you understand!' seems to be the meta-hermeneutic message of the psychoanalytic interpreter who presents himself, as Jacques Lacan famously put it, as the 'subject supposed to know' (*le sujet supposé savoir*). Since Freud himself held to the idea that 'understanding and cure almost coincide',[4] it would pay to look closely not just at how the analyst was supposed to proceed in therapy, but also what the underlying theoretical structure and presuppositions of the hermeneutic method are.[5] Although we cannot dispute that psychoanalytic interpretations can be perfectly *justified* in the light of Freudian theories and that there are interesting social, cultural and perhaps even political constraints on what will count as an 'acceptable' psychoanalytic interpretation within a community, it does not follow that they adequately *explain* the interpreted phenomena. What accounts for this intriguing phenomenon and what disguises the illicit move from justification to explanation?

A final set of questions concerns crucial liaisons between the hermeneutic structure of the psychoanalytic method and striking *social* features of the (Freudian) psychoanalytic edifice. Critical historians of psychoanalysis have tried to link the enormous success of the discipline not only with the particular therapeutic lacuna Freud discerned, but also with the rapid emergence of societies that held Freud as their undisputed master (Roazen, 1975; Breger, 2000; Borch-Jacobsen and Shamdasani, 2006). In 1902 Freud founded the *Wednesday Society* which rapidly grew and was renamed in 1908 as *The Vienna Psychoanalytic Society*. Other psychoanalytic societies quickly followed. But what was striking, as many pointed out, were two distinctive traits that would set apart the psychoanalytic movement from other scientific societies: first, there was—as many of Freud's contemporaries testified—the almost religious atmosphere, and, secondly, the intolerance of dissent and opposition in public or print. Recalling the early Wednesday evening meeting, Max Graf (father of Freud's famous child patient, 'Little Hans') wrote that "there was an atmosphere of the foundation of a religion in that room [...] Freud Pupils were his apostles [...] Good-hearted and considerate though he was in private life, Freud was hard and relentless in the

---

[3] Norman Holland, cited in Olsen (1986, p. 204).
[4] Cf. Freud (1933, p. 145).
[5] Cf. Roazen (1975, pp. 131).

presentation of his ideas" (Max Graf, quoted in Roazen, 1975, p. 193). Freud insisted on absolute loyalty (Roazen, 1975, p. 308). The hermeneutic framework was embedded in a movement with distinctively cult-like traits, closely controlled by a charismatic leader who was surrounded by disciples, "like the Paladins of Charlemagne, to guard the kingdom and policy of their master" as Breger (2000, p. 209) colourfully put it. While many have considered this an idiosyncratic (and perhaps deplorable) trait of Freud as Founding Father, we suggest there are deeper explanatory links between the quasi-religious character of the movement, Freud's opposition to criticism and the very existence of psychoanalytic facts. Rather than being an accidental feature of the theory and the method, the social features of psychoanalysis are (we argue) crucial for the introduction and maintenance of the psychoanalytic facts Freud 'discovered'. Every psychoanalytic school has in some way or another developed distinctive sect-like traits, and the fractioned history of psychoanalysis is mainly one of competing societies and contested claims of orthodoxy (Breger, 2000; Borch-Jacobsen and Shamdasani, 2006).

2 Unintended institutional facts

Our explanatory hypothesis involves an application of John Searle's theory of institutional facts (Searle, 1995; Searle, 2010; Lagerspetz, 2006). The explanatory strategy rests on an inference to the best (because unifying) explanation of the well known and independently confirmed phenomena described in section 1: how can the following, *prima facie* unconnected phenomena be given a unified explanation: the capacity to understand everything, the epistemically closed character of the theory, the difficulty of integrating psychoanalytic findings in other disciplines and the remarkable social structure of the psychoanalytic movement? All these features suggest that the key hermeneutic claims of Freudian psychoanalysis help create *institutional facts*, in the precise sense developed by John Searle in *The Construction of Social Reality* (1995) and, more recently, *Making the Social World* (2010).

Searle's theory of institutional facts builds on insightful suggestions provided by Elisabeth Anscombe (1958), and was further explored and modified by philosophers like Tuomela (2002) and Lagerspetz (2006), among others. While Freud himself consistently presented his findings as *natural* facts and psychoanalysis as a science that discovers those facts, we hold that the Freudian interpretations operates successfully in virtue of the unintended creation and maintenance of *institutional* facts, in a precise technical sense to be explained below. The continued existence of those institutional facts requires the creation of a complex pattern of shared beliefs among the 'believers,' who contribute further 'confirmations' of the truth of the claims instigated by Freud. This explains why psychoanalysis functions as a tightly controlled thought system, with its various schools uniformly characterized by strong hierarchical relations, absolute loyalty to the master/founder, and the expulsion of dissidents. The institutional facts have the function of making sense of you (your dreams, actions, mishaps, etc.), they create obligations and permissions, determine whether you are psychologically healthy or not, etc.

This assessment of the outcomes of the psychoanalytic method doesn't rest on a global constructivist approach to science or hermeneutics. Social constructivism provides an implausible account of science and must be rejected on independent grounds (Kukla, 2000; Boghossian, 2006, and cf. Boudry and Buekens, 2011, for a critique of social constructivism in science and psychoanalysis).[6] In the conclusion, we shall briefly come back to this issue and explain why a psychoanalytic pronouncement but not, say, the postulation of a particle in physics, can result in in the creation of an institutional fact.

Let's begin with an observation about the structure of a psychoanalytic interpretation. Since psychoanalytic interpretations are almost instantly recognizable, it suffices to introduce simple examples to illustrate their remarkably surface structure: a relatively obvious anthropological phenomenon X is assigned a distinctive psychoanalytic property Y: [7]

> Agoraphobia in women (X) is the repression of the intention to take the first man one meets in the street (Y). (Freud in Masson, 1985, p. 217–18)
>
> The deepest unconscious root of anti-semitism (X) derives from the castration complex (Y). (*SE X*, p. 36)
>
> Early man's control of fire (X) derives from the renunciation of the 'homosexually-tinged desire to put it out with a stream of urine. (Y) (*SE XXII*, p. 187)

Given any able interpreter's competence to construct narrative justifications that link these concrete psychoanalytic identifications with theoretical claims of Freud (or Lacan, or...), the theoretical claims come to justify the redescriptions of phenomenon X as psychoanalytic phenomenon Y. In 2005, cultural theorist Jerry Flieger put the critic of Freud in the X-position and obtained the following unsurprising result:

> Indeed, the vitriolic attacks (on psychoanalysis) (= X) may in fact be considered in terms that Freud himself contributed to cultural

---
[6]Cf. Stern (1992) for a defense of a social constructivist approach to psychoanalysis. We briefly criticize global social constructivism in the last section of this paper and in Boudry and Buekens (2011).

[7]Examples drawn from Esterson (1993, p. 244).

discourse—as examples of classic denial (= Y_1) (Freud is not my intellectual father), paranoid generalization (= Y_2) (Freud is to blame for everything), or intellectual hysteria (= Y_3) (Freud reduces everything to sex). (Flieger, 2005, p. 9, our additions)

The natural question to ask here is whether a perfectly *justified* interpretation of X as psychoanalytic phenomenon Y (and supported by a suitable narrative that connects central claims with the phenomena interpreted) *explains* phenomenon X. We'll return to this intriguing question in the final section of this paper. For now, we wish to remark on a first intriguing feature of these identifications, viz. the fact that they are framed in terms of what Clifford Geertz (1983) calls *experience-distant concepts*, which contrast with *experience-near concepts*:

> An experience-near concept is, roughly, one that someone—a patient, a subject—might himself naturally and effortlessly use to define what he or his fellows see, feel, think, imagine, and so on, and which he could readily understand when similarly applied by others. An experience-distant concept is one that specialists of one sort or another—an analyst, an experimenter, an ethnographer, even a priest or an ideologist—employ to forward their scientific, philosophical, or practical aims. 'Love' is an experience-near concept, 'object cathexis' is an experience-distant one. 'Social stratification' and for most people in the world even 'religion' (and certainly 'religious system') are experience-distant; 'caste' and 'nirvana' are experience-near, at least for Hindus and Buddhists...(Geertz, 1983, p. 57)

David Jopling, who draws our attention to this distinction, adds that experience-distant concepts in psychodynamic therapy

> include concepts such as unconscious forces, resistance, repression, denial, regression, transference, reaction formation, reversal, sublimation and splitting. These concepts come to play a central role in interpretation and insights. Clients learn to think of themselves in terms of these new concepts, so much so that what they first encounter as an experience-distant concept upon first entering treatment may evolve in an experience-near concept. (Jopling, 2008, p. xxiv)

For our purposes, it is important to draw attention to another feature of Geertz' distinction: the experience-distant concepts that figure in interpretations do not come with independent criteria for what will count as evidence for the truth of claims or interpretations in which they occur. In other words, outside the theory which *defines* what counts as evidence in terms of other experience-distant concepts, no independent application criteria exist. The theory introduces the concepts, and the interpretations rest on accepting the theory to the extent that—as Joplin rightly points out—the original experience-distant concept becomes an experience-near concept. Trained psychoanalysts and psychonalytic interpreteres characteristically 'see' the Freudian meanings in what they interpret.[8]

Although the experience-distant character of the central concepts in psychoanalysis is important, this feature by itself does not have much explanatory power. The hidden power of experience-distant concepts comes to light when we combine the observation with the characteristic speech act in which these concepts figure. The characteristic outcome of a psychoanalytic interpretation is a proposition of the form 'X is, turns out to be, or should be, identified with psychoanalytic phenomenon Y', where the Y-position is occupied by an experience-distant psychoanalytic term (or cluster of terms) derived from the psychoanalytic theory that forms the background theory. The identifications presented are based on carefully selected contextual evidence unearthed by the psychoanalytic technique, but which, on closer inspection, turn out to be just more psychoanalytic identifications of the same type. Thus, dream symbols are presented as evidence for the identification of dreams (X) as wish-fulfilments (Y), but the symbols themselves are X-components within propositions of the form 'X is a symbol for/stands for Y', where Y designates a psychoanalytic concept. More importantly, however, the psychoanalytic identifications are presented as contents of *descriptive* speech acts. These speech acts often take the form of assertions, but can also sometimes be hypotheses or conjectures. This suggests that they are based on empirical inquiry, and are therefore put forward as refutable statements and inductive generalizations. If the psychoanalytic assertion or conjecture is true and justified, we gain psychoanalytic understanding. This brings us to the second element in our account of psychoanalytic facts: their *declarative origin*.

Propositions—here uncontroversially thought of as contents of assertions—can be objects of different types of speech acts (Searle, 1969). When used in assertions, they are contents of speech acts with a mind-world direction of fit—they are true if, and only if, the world is as the assertor claims it to be (Searle, 1969; Humberstone, 1992). But propositions can also function as the content of declarative speech acts which state that X from now on counts as Y, as in the declarative statement that this (the referent of the X-term) is (i.e., will from now on count as) that (the referent of the Y term, e.g., money, property, a border, or a valid contract) (Searle, 1995, 2010; Smith, 2003).[9] Used in descriptive speech acts of which assertions are the prime example, the proposition that X is Y describes a fact or phe-

---

[8]The characteristically 'vague' and 'open' character of key psychoanalytic concepts and its role in the effectiveness of psychoanalytic interpretations is further examined in Borch-Jacobsen (2005) and Cioffi (1998).

[9]We assume that contextual conditions for successfully creating an institutional fact are satisfied.

nomenon; however, used in declarative speech acts, the proposition helps creating a new institutional fact—the speech act has both a mind-word and world-mind direction of fit.[10] For example, uttering 'Bedtime now!' (said to children at around 8 pm) creates a miniature institutional fact (the time to go to bed for the children). The real or fictional declarative 'This is (i.e., from now on counts as) one euro' created a new monetary unit in Euroland (which is, by the way, also an institutional entity).[11] The creation of institutional facts via declaratives requires that the speaker—the person who issues the declarative—have the relevant authority to issue the declarative. Moreover, possessing this authority requires the occupation of a specific institutional role, in order to successfully declare that X counts as Y (only the U.S. President can appoint a member of the Supreme Court).

Declarative language use with the purpose of intentionally creating institutional facts is part and parcel of our speech act repertoire and to a large extent responsible for the wealth of *bona fide* institutional facts that surround us: money, property, borders, contracts, world records, tenured professors and enrolled students. Of course, as Searle (1995, p. 47) points out, not all institutional facts are *explicitly* and *consciously* introduced by declaratives. For our purposes, it is sufficient that some institutional facts are introduced in this manner, and that all institutional facts *could* be so introduced. Institutional facts should be contrasted with brute or natural facts that exist independently from our attitudes directed at them (Anscombe, 1958). Many natural facts (objects, properties) are discoverable by us, and can become bearers of institutional properties, whereas institutional facts are (explicitly or implicitly) created by us and are of an abstract nature. 'Being worth one euro' is not an observable property, although our observations of many objects, events and properties are inevitably laden with concepts derived from descriptions under which they are institutional facts ('I *see* that this coin is worth one euro', or 'I *am witnessing* Tom and Jerry's marriage'.). Lagerspetz (1989, p. 9) points out that "terms which are used to refer to institutions are in some sense like theoretical terms", and Barry Smith (2003) makes the additional point, consistent with Lagerspetz' observation, that the mass of social facts that surround us form 'a huge, invisible ontology' (Smith, 2003, p. 17).

Institutional facts *introduced* by (explicit or implicit) declaratives *continue* to exist or are maintained (Searle, 2010, p. 102) only if very specific attitudes in the introducing agent and his intended audience are present or can be induced (cf. also Pettit, 1993; Searle, 1995; Tuomela, 2002; and

---

[10] Our talk of facts and phenomena is intended to leave open the ontological status of facts (true statements or truth makers). Nothing in our discussion or in Searle's account of institutional facts depends on this point.

[11] The 'counts as' locution makes the declarative character of the speech act explicit.

Lagerspetz, 2006, whose account we shall follow in this paragraph).[12] Their *Performativity* (as a feature of institutional facts) is based on a shared attitude toward the institutional fact, and may contribute to the truth of a sentence describing the fact. The stock example is money: if a group accepts that certain pieces of metal count as money, then, under the appropriate circumstances, these objects *are* specimens of money for that group. Accepting that certain objects count as money by members of a group is not based on independent evidence that the coins serve as money: rather, the shared attitude creates the institutional facts, which certain sentences then go on to describe correctly. The second feature of institutional facts is *Reflexivity*: if a sentence describing the institutional fact is true, the relevant attitude must be present in the speaker (Searle, 1995, p. 32–4; Tuomela, 2002). The third feature is *Qualified Realism*: institutional terms refer to real (abstract) objects and properties; their referents are not fictional entities and they are not intended as fictions. Statements about money, interest rates, property, borders or—as we shall argue—distinctively psychoanalytic entities like *penis envy*, a dream's *latent content*, or *sublimation of the sex drive*—can thus be literally true or false. But the objects whose truth and falsity they represent do not exist independently of all representation. A global *error theory* about institutional facts doesn't give an adequate account of our intuitions about institutional facts (Searle, 1995, p. 90ff).

This brief sketch of the cognitive make-up of a group accepting pieces of paper as money or a river or virtual line as a border, doesn't imply that its members need to explicitly know, or need to be explicit about the fact that that they are maintaining the existence of (explicitly or implicitly introduced) institutional facts. In other words, the threefold structure that guarantees the continued existence of a (system of) institutional fact(s) need not be seen or recognized by the participating group as necessary conditions for what they *take* to exist. Moreover, they need not be aware of the role of their shared, coordinated beliefs as contributing to the truth of sentences describing those facts. It is, however, *possible* for them to come to realize that what they take to be money or a border is an institutional fact, and they can also come to realize that an implicit or explicit declarative of the form 'X counts as Y' lies at the institutional fact's origin, and that what they took to be a 'brute' or natural fact turns out to be an institutional fact. (In this sense, the conceptual distinction proposed by Searle and others is a kind of Wittgensteinian reminder of what we knew all along, albeit implicitly).

What Searle (1995) calls the *function* assigned to X in the formula 'X counts as Y' cannot be performed solely in virtue of X's physical properties;

---

[12] Of course, these authors do not agree on all the details about how institutional facts are generated, maintained or go out of existence. For the purposes of this paper, we shall use Lagerspetz' useful account of performativity.

it requires "our agreement or acceptance that it [the function, FB&MB] can be performed" (Searle, 1995, p. 42). When the archaeologist speculates that a piece of metal found on site functioned as money, he is making delicate assumptions about quite specific shared beliefs (and attitudes) in a former society. Of course, the particular artifactual shape of the piece of metal or the place where it was found is excellent evidence for the hypothesis that the pieces were artifactual, but the set of physical properties will never be *sufficient* to qualify it as money. Thus, although institutional facts and the embedding institutional frameworks supervene on the physical properties of their realizers, the concepts required to account for them *qua* institutional facts (i.e., concepts involving shared beliefs) are not reducible to physical concepts. Since institutional facts are not reducible to collections of brute facts, an institutional framework is 'closed' in two directions: there are no non-mental properties that are necessary and sufficient for something to count as this or that institutional fact; and, conversely, an entity's institutional status is never evidence for the obtaining of non-institutional properties of its concrete, material bearers (indeed, it might even be argued that some institutional facts have no material bearers at all—cf. Smith (2003) for further discussion). This doesn't exclude the fact that institutional facts have intended and non-intended non-institutional *causal* consequences (excessive inflation can cause poverty and famine), and the creation of institutional facts sometimes requires substantial modifications of natural facts in order for the latter to be manageable realizers of institutional properties.[13]

The final step in our explanatory model is that a group can come to realize that they have maintained a system of institutional facts, which were mistakenly qualified as natural facts. Searle makes the following observations:

> The process of the creation of institutional facts may proceed without the participants being conscious that it is happening according to this form... (Searle, 1995, p. 47)

> Most of these things (the creation of institutional facts, FB & MB) develop quite unconsciously, and indeed people typically are not even aware of the structure of institutional reality. It often works best when they have false beliefs about it. So there are a lot of people in the United States who still believe that a dollar is only really money because it is backed by all that gold in Fort Knox. This is total fantasy, of course. The gold has nothing to do with it. And people hold other false beliefs. They believe someone is king only because he

---

[13] Cf. Smith (2003) for further discussion. It is possible that the X-position is occupied by an institutional fact and that the Y-position assigns a second-order institutional fact. E.g., to become President of the U.S., one must be a U.S. citizen.

is divinely inspired, or even believe that marriages have been made by God in heaven, and so on. I am not trying to discourage them *because often the institution functions best when people hold false beliefs about it.* (Searle, 2001, p. 37–38; our italics)

Note, first, that Searle holds that false beliefs about the nature of a practice can be useful for the continued existence of the practice, which, as Lagerspetz (2006) correctly points out, falls short of admitting their necessity for conceptual and/or practical reasons (A community *can* be fully aware of the fact that they're involved in the creation of specific institutional facts.). The situation is even more complex than Searle suggests: an unintentionally created institutional fact can explicitly and consciously be taken to be a natural fact, which may later be revealed ('unmasked', one is inclined to say here) as a ('mere') institutional fact. The latter point allows for the possibility that an unintended institutional fact, and its embedding practice, can become an *intended* and *explicitly recognized* institutional fact or practice. When an unintended institutional fact 'survives' after being explicitly revealed as institutional fact, its function can often be improved and further useful extensions of the practice can be introduced. On the other hand, when it becomes common knowledge among participants that a practice, explicitly presented as involving natural facts, was based on the creation and maintenance of institutional facts, continuation of the practice will be problematic: it will either die out, be abandoned, or even be explicitly rejected.

We shall now show that the cluster of phenomena that Searle describes in the last quote—namely, that people typically are not aware of the structure of the institutional reality that they are involved in; and, that an institution sometimes functions best when people hold false beliefs about it (although this should not be generalized!)—help to characterize the psychoanalytic practice as a distinctively hermeneutic practice. Freud's central theoretical identifications were *declaratives*, presented as *assertions*. The Freudian declaratives served to introduce a complex system of institutional facts which, when accepted by others, came to be known as the central psychoanalytic truths. So, while Freud himself was responsible for their introduction, he alone could not *maintain* the continued existence of these facts. Their continued existence required that others contribute to the truth of sentences describing the fact by accepting what Freud declared to be the case. In other words, it was necessary for others to come to believe that which Freud thought he merely described but in fact unintentionally created. That is, to put it bluntly, why there *had* to be a psychoanalytic movement.

Let's pause here to explain the 'unintended' character of institutional facts. The first thing to notice is that, as Sally Haslanger (2006) succinctly puts it in a related analysis examining the social construction of concepts like *race* and *gender*, 'our meanings are not always transparent to us'. What was not immediately transparent for Freud, and his intended audience, was the specific nature of the speech act he performed when he said that human phenomenon X is—or *counts as*—psychoanalytic phenomenon Y. While he explicitly maintained that he was merely *describing* facts, he was, in fact, creating new entities within an unfolding institutional framework. This particular form of non-transparency regarding the force of institutional statements is made possible because the proposition *that X is Y* can also be used in speech acts with descriptive force (mind-world direction of fit). We contend that it is the declarative character of the original Freudian psychoanalytic speech acts—the ones that initiated the institutional facts—that was neither transparent to Freud, nor to the psychoanalytic interpreter who applies Freudian theories in his hermeneutic practice. Freud's immensely important and effective 'factive rhetoric'—typified by his insistence on having *discovered* the meaning of dreams, his description of the technique as *unearthing* the psychoanalytic meaning of symptoms and dreams, and the analogies with archaeology and puzzle-solving (cf. below; p. 48)—played a key role in hiding from view the phenomenon we just described.[14]

The institutional character of psychoanalytic facts becomes even more plausible when one takes a closer look at the social acceptance-conditions responsible for their continued existence or *maintenance*, to use Searle's concept (Searle, 2010, p. 102). As mentioned before, psychoanalysis' perspicuous institutional character includes the striking cult-like character of the enterprise, its strongly hierarchical internal organization, and the felt urge to disseminate psychoanalysis as if it were a religion. All these features, well documented by the historians of psychoanalysis, eminently enhanced the carefully controlled distribution of the beliefs necessary to contribute to the existence of what made the sentences describing psychoanalytic facts true (cf. below; p. 56: the *performativity* feature). The 'cult-like' structure of psychoanalysis turns out to be a predictable consequence of the fact that a system of institutional facts cannot be maintained on the basis of empirically-founded beliefs, or be supported by independent evidence. *Any* viable system of institutional facts requires the careful steering and coordination of the supporting beliefs held by the collective responsible for their continued existence. Since no independent evidence supports these beliefs—as was already noted early on in the history of psychoanalysis and later explicitly argued by critics of psychoanalysis (cf. above; p. 35)—various

---

[14]Cf. also Esterson (1993, p. 205ff.) for a further analysis of Freud's claim that he was only communicating material based on 'clinical observations'. As Esterson points out, "the frequent references to his 'findings' and 'discoveries' inevitably creates, in the mind of the reader, a feeling that there must be at least *some* substance to it" (Esterson, 1993, p. 206).

non-epistemic strategies had to be introduced to stabilize the beliefs that contribute to the truth of sentences describing psychoanalytic facts. Faithful followers apply the central declaratives and 'find new confirmations'; committees ensure that the central declaratives are not modified; dissident figures likely to undermine the continued existence of the institutional facts—by modifying or rejecting the central declaratives—must be renounced. We leave it to the reader to re-read the intriguing and well-documented features of the sociology of the psychoanalytic culture as further extensions and applications of our institutional account of psychoanalytic facts. In the next section we connect the abstract character of institutional facts with the closed character of psychoanalytic theories. Our favourite example is Freud's brilliant introduction of his best known institutional fact: that dreams are wish-fulfilments.

3   The invisible ontology of the wish-fulfilling dream

In *The Interpretation of Dreams* (published in 1899), Freud made it immediately clear that he was breaking with the past by distinguishing the 'manifest' dream and its 'latent' dream-content. What is often described as his 'basic hypothesis' can be roughly summarized as follows: a dream is produced when an unacceptable, repressed infantile wish becomes active during sleep. When the unacceptable wish threatens to break through into consciousness, a process that safeguards sleep intervenes. This transforming process, the 'dream work', distorts the wish so that it appears in the dream in a disguised form. The analyst is capable of interpreting the dream by following the associations of the dreamer, and by applying his knowledge of the dream symbols.

The relevant question that arises in the context of our inference to the best explanation is whether these are empirical hypotheses, or constitute a carefully composed set of declaratives (the central declarative being 'dreams are wish fulfilments') which—when duly accepted and systemically maintained by participants—*create* Freud's necessarily invisible (abstract) ontology of dreams. Interestingly, Freud himself made ambivalent claims about their status. He often presented his central claims as hypotheses, supported by 'clinical observations' (or so he described his own interpretations of his dreams and those of his patients). If readers accepted the evidence, they themselves were expected to adopt the same epistemic attitude of believing the interpretation. On the other hand, Freud inadvertently stressed a key feature we have assigned to institutional facts: no *evidence* is required to create the central institutional fact, here disguised as a 'postulate'. Consider the subtle rhetoric with which Freud draws his readers into acceptance of his basic premises, as illustrated in the first lecture of the *New Introductory Lectures* (published in 1933):

> We have—quite arbitrary, it must be admitted—made the assumption, adopted as a postulate, that even this unintelligible dream must be a fully valid psychical act, with sense and worth, which we can use in analysis like any other communication. Only the outcome of our experiment can show whether we are right. If we succeed in turning the dream into an utterance of value of that kind, we shall evidently have a prospect of learning something new and of receiving communication of a sort which would otherwise be inaccessible to us. (*SE XXII*, pp. 8–9)

That dreams are fully valid psychical acts—that is, wish-fulfilments—is 'quite arbitrary' and 'adopted as a postulate'. But given this postulate, the interpretation of a particular dream *as* a wish fulfilment cannot represent an attempt to find evidence for a hypothesis (as what happens in a genuine experiment). Rather, the postulate presents a hermeneutic challenge to the analyst. The challenge amounts to creating an interpretation that is *consistent* with the central institutional fact that dreams are wish fulfilments. Neither proof nor refutation of the declarative statement is possible, because there is no such thing as *independent* evidence for an institutional fact.

Having declared that dreams are wish-fulfilments, Freud could create more specific institutional facts consistent with the initial postulate, but present them as assertions whose content would constitute *evidence* for the postulate itself. The core structure of the *Traumdeutung* thus turns out to be a brilliantly woven network of institutional facts: dreams (X) have (count as having) manifest content and latent content (Y). The unconscious wish is its latent content, therefore X's manifest content is (counts as being) censured by the unconscious (Y). Objects, situations, or persons occurring in the manifest content (X) are (count as) symbols for often sexually laden acts or objects (Y). The analyst can freely select from the patient's associations to create a narrative in which the internally consistent system of declaratives appear as a coherent interpretative description of the patient's vicissitudes. Wittgenstein's suggestion that dreams might as well be interpreted as expressing unconscious fears, rather than desires, illustrates the arbitrary nature of Freud's initial declarative (Barrett, 1967). Wittgenstein's critical observation unintentionally reveals another key feature of the institutional character of Freudian psychoanalysis: if dreams merely count as manifestations of unconscious wishes, alternative frameworks could be developed in which dreams would count as manifestations of unconscious fears, and reveal the initial 'postulate' as a merely arbitrary fact. Wittgenstein's comment suggests that he too must have sensed the institutional character of the Freudian edifice. Freud's conclusion, near the end of chapter 4 of the *Traumdeutung*, that "a dream is the (disguised) fulfilment of a (suppressed or repressed) wish", is not only presented as the key to interpreting all dreams (a description that is akin to his belief that the Oedipus complex is 'universal'), but as the central institutional fact, with which interpretations had to cohere. The truth about the unconscious as revealed in dreams was *created* by Freud and *maintained* by his followers.

Hiding from view the institutional character of the wish-fulfilling dream is enhanced by Freudian comparisons and analogies that strongly suggest that latent content is out there, to be found (and not constructed) by the psychoanalyst. Freud's famous puzzle-metaphor eminently served this purpose:

> I have a picture-puzzle, a rebus in front of me. It depicts a house with a boat on its roof, a single letter of the alphabet, the figure of a running man whose head has been conjured away, and so on. Now I might be misled into raising objections and declaring that the picture as a whole and its components parts are nonsensical. A boat has no business to be on the roof of a house, and a headless man cannot run [...] but obviously we can only form a proper judgement of the rebus if we put aside criticisms such as these of the whole composition and its parts and if, instead, we try to replace each separate element in some way or other. The words which are put together in this way are no longer nonsensical but may form a poetic phrase of the greatest beauty and significance. A dream is a picture puzzle of this sort. (*SE IV*, p. 277–8)

Rebus-metaphors and archaeological analogies ('*saxa loquuntur!*'; *SE III*, p. 192) effectively concealed from Freud and his followers how the technique inevitably resulted in the unintended creation of new institutional facts—the interpretations of dreams—rather than the discovery of natural phenomena. The 'free associations' generated on Freud's couch provided a rich source of personal anecdotes and factoids to construct narratives in which these institutional facts could appear as natural facts. Unsurprisingly, there are no dream-symbols in Freud's theory that are inconsistent with the dream as wish-fulfilment (the core institutional fact). Interestingly enough, the puzzle-metaphor had already figured in the notorious *Seduction Theory* (Freud, c. 1896). Regarding the 'scenes' that had to be remembered by his patients, Freud wrote:

> It is exactly like putting together a child's picture-puzzle: after many attempts we become absolutely certain in the end which piece belongs in the empty gap; for only that one piece fills out the picture and at the same time allows its regular edges to be fitted into the edges of the other pieces in such a manner as to leave no free space and to entail no overlapping. In the same way, the contents of the infantile scenes turn out to be indispensable supplements to the associative and logical framework of the neurosis, whose insertion makes its course of

development for the first time evident, or even, as we might often say, self-evident. (*SE III*, p. 205)

By postulating the infantile scenes as "indispensable supplements to the associative and logical framework of the neurosis", it seems as though Freud was putting forward a hypothesis based on an inference to the best explanation. However, one might as well claim that Freud *declares* what the neurotic symptoms of his patients had to count as.[15] Insofar as Freud's patients came to believe his interpretations were empirically adequate accounts of neurotic phenomena, they could benefit from the therapy. This phenomenon eminently illustrates Searle's earlier observation that many people may hold false beliefs about institutional facts or practices in which they engage ("the institution often functions best when people hold false beliefs about it"—cf. above; p. 44). In Freud's case, the false belief was that only empirical claims and testable theories about natural psychic phenomena were put forward.[16] On the other hand, institutional facts may have causal consequences in those whose beliefs help maintain those facts—hence the real (and sometimes beneficial) effect of the Freudian interpretation on his patients.

In a famous letter to his one time intellectual ally Wilhelm Fliess dated September 21, 1897, Freud introduced what arguably turned out to be the central institutional fact that later came to characterize classical Freudian psychoanalysis: neuroses find their origin in infantile sexual fantasies. Once this declarative began to be accepted by others (first his patients, then a small circle of adherents, and later a whole network of psychoanalytically minded doctors, psychiatrists and lay analysts), neuroses (X) *became* manifestations of infantile sexual phantasies (Y). From 1899 onwards, dreams became wish-fulfilments. Similarly, in 1902, slips of the tongue obtained the status of symptoms of unconscious drives, desires, or beliefs and consequently, the patients' free associations became revelatory of his or her unconscious. In *Drei Abhandlungen zur Sexualtheorie* (1905) sensual acts of babies and toddlers (X) became manifestations of the oral, anal and genital phases (Y), a controversial institutional fact presented as evidence for—and consistent with—another institutional fact: the Freudian libidinal drive. Carl Gustav Jung later introduced the collective unconscious, and Alfred Adler the inferiority complex. In 1924, Otto Rank 'discovered' the trauma

---

[15]Freud's approach in the *Etiology of Hysteria* is now widely seen as epistemically flawed, primarily because Freud himself suggested the disturbing scenes to his patients (Esterson, 1993).

[16]The fascinating question is, of course, what happens when the facts that Freud described as natural facts are exposed as institutional facts. This raises the further question of whether the therapy continues to engender its positive (but placebogenic) effects if no one believes that the therapist is uncovering natural phenomena. Cf. Jopling (2008) for an excellent and up to date discussion of psychic placebos.

of birth, an institutional fact which proved to be inconsistent with Freud's own framework. When Ernest Jones, Freud's first biographer, pointed out that Freud couldn't find any evidence at all for the trauma of birth, Freud understandably interpreted Rank's claim as an empirical hypothesis. Had he himself accepted that dreams and symptoms might simply *count as* manifestations of the Rankian trauma of birth, he would surely have accepted, as evidence, whatever further institutional facts Rank presented in support of *his* central institutional fact.

Once Freud *declared* that dreams are/count as wish-fulfilments and his declaratives were duly accepted by others (e.g., by the *Wednesday Society* in Vienna, the first followers in Zürich (Karl-Gustav Jung, Eugen Bleuler), the audiences attending his lectures, and his patients), the institutional fact-generating declaratives concerning wish-fulfilling dreams—which Freud and others had made—became *literally true* (recall our *qualified realism* about institutional facts—cf. above; p. 43). Stressing the evidence-based nature of dream interpretations engendered the belief that what he had 'uncovered' were *empirical* facts; but acceptance unavoidably helped to maintain the (unintended) institutional fact that equated dreams with wish-fulfilments. Contrary to Freud's own suggestion that neurology would eventually confirm the existence of postulated mechanisms like repression and sublimation, his ontology had to remain *abstract* and *invisible* (cf. above; p. 41). The Freudian unconscious and its ingredients are a complex system of institutional facts, which are by definition abstract entities. (We'll never find evidence for *abstract, institutional* facts in soggy grey matter.). It comes as no surprise that empirical evidence for the *Freudian* unconscious has never been discovered. (No one should be tempted to confuse the ingredients of the *Freudian* unconscious with perfectly empirical facts about sub-personal processes in the brain.)

Earlier, we pointed to the closed character of psychoanalysis: psychoanalytic 'findings' cannot be used as evidence in other disciplines or fields of study, and the psychoanalytic technique never became part of the repertoire of *bona fide* scientific or hermeneutic methods in other disciplines. But there is another phenomenon: the distinctive feeling (experienced by those who apply the theory) that a psychoanalytic interpretation of a phenomenon X marks the end of inquiry, and further suggests (at least for those who accept it) that no additional work needs to be done to understand the phenomena occupying the X-position. 'Now I understand!' is the feeling when a phenomenon X is placed under a psychoanalytic label Y. In fact, no additional work *can* be done, for the resulting identification, an institutional fact, will never count as evidence outside the field of psychoanalytic inquiry. Furthermore, no Freudian insight will be *epistemically* supported by facts outside the realm of psychoanalytic institutional facts. The latter aspect of the closed character of the theory follows from the general observation that extra-institutional facts never confirm the existence of, or provide evidence for, institutional facts. Notice however, that the closed character of psychoanalysis qua institutional system still allows that the psychoanalytic practice may have extra-psychoanalytic *causal* consequences (e.g., when used in healing practices). While it is certainly true that the central psychoanalytic concepts became *experience-near concepts* for those who accepted the theory (cf. Clifford Geertz' distinction, explicated on p. 39), it does not follow from this phenomenon that the central claims became more empirically verifiable. A better explanation would be to see the shift from experience-distant to experience-near concepts as an unavoidable *cognitive* side-effect of accepting and maintaining Freudian institutional facts. Those who accepted Freud's pronouncements not only helped in creating and maintaining psychoanalytic facts; they came to think of countless phenomena in terms of the psychoanalytic concepts that functioned as central ingredients of the theory.

4   Discussion: illusions of understanding, and a comparison with social constructivism

Our assessment of psychoanalysis as a system of carefully maintained institutional facts connects some well-known independent objections voiced by, among others, Karl Popper and Adolf Grünbaum. Popper's critique of psychoanalysis as a pseudoscience inspired philosophers like Paul Ricoeur and Jürgen Habermas to requalify psychoanalysis as a hermeneutic method, and they dismissed Freud's own positivistic interpretation of his theories as a *Selbstmissverständnis*. An intriguing argument, in defense of their approach, has been put forward by—among others—philosophers like Jim Hopkins, Marcia Cavell, Thomas Nagel and Richard Wollheim. According to them, psychoanalysis' key concepts and principles should best be seen as *non-conservative extensions of our folk psychology*—the ineliminable hermeneutic practice *par excellence* that helps us to better understand ourselves and others (Carruthers and Smith, 1996; Hutto and Ratcliffe, 2007).[17] As Winter points out, "(t)he cultural prominence of psychoanalysis has arisen from the persuasiveness of that thought style beyond the confines of professional psychotherapeutic practice, so that psychoanalytic knowledge also appears in the guise of a popular-cultural approach to self-understanding" (Winter, 1999, p. 17; and compare Kusch, 1999). Folk psychology we define here as

---

[17]'Ineliminable' because we reject the reduction of its key concepts to neurological concepts. Cf. also Footnote 2. The 'extension view' has at least two versions: those who see the extension as primarily formal (psychoanalytic explanations as similar in type to commonsense explanation), and those who see the extension as substantive, that is, as extending the realm of beliefs, desires and emotions.

a complex cluster of meta-representational concepts and constitutive principles, which are acquired at a very young age. These elements purportedly make it possible for us to understand each other, by ascribing mental states such as beliefs, desires, intentions, emotions and intentional actions via rationalizing actions and empathy. Our key objection to the hermeneutic counter-move is that folk psychology makes sense of distinctively *natural* phenomena (beliefs, desires, actions and emotions, cognitive capacities to empathize with others). The Freudian hermeneutic technique, on the other hand, introduces artificial concepts and creates, via declaratives accepted by the psychoanalytic community, the very phenomena it claims to explain. One cannot therefore argue that it is because a psychoanalytic concept can be assigned to a certain phenomenon (and that this designation can be perfectly *justified* in terms of the background theory) that psychoanalytic interpretations yield genuine *understanding* as the application of natural folk psychological concepts do. This is consistent with the observation that applications of psychoanalytic concepts in a therapeutic setting can produce beneficial effects for patients: the Freudian therapy, like many other psychodynamic therapies, has a significant placebogenic effect (Jopling, 2008). Our analysis thus suggests that a more fundamental *Selbstmissverständnis* lies at the heart of psychoanalysis: what the method explicitly presents as interpretations of natural meaningful phenomena turns out to rest on the unintended creation of institutional facts. There is an important sense in which Freud never quite understood what he did. The construction of the institutional context that made this *Selbstmissverständnis* possible was a genuine *tour de force*: on the one hand, Freud had to present his claims as hypotheses backed up by epistemic reasons; on the other hand, given the obvious weaknesses of his arguments and evidence, and the quickly acknowledged arbitrary nature of his interpretations by his critics, he had to create a community of adherents who accepted his pronouncements to maintain the existence of the facts he unintentionally created.

Why is it so tempting to think that we understand human or cultural phenomenon X 'better' when presented with its psychoanalytic interpretation? A psychoanalytic interpretation links theoretical claims of Freud with re-descriptions of concrete human phenomena. The illusion of thereby having understood, that is, having *explained* a particular phenomenon, derives from a subtle confusion between justifications (a relation between beliefs) and explanations (a relation between facts). Carl Hempel (1965) has pointed out that it doesn't follow from the fact that 'X's belief that p justifies his belief that q', that '(the fact that) p explains why q is the case'. My belief that the thermometer is sinking today surely *justifies* my belief that it will be cold tomorrow, but the movements on the scale don't *explain* why it's going to be cold tomorrow (that fact is ultimately explained by meteorology and physics). Justifications provide reasons for *believing* that p while explanations yield understanding of the *fact* that p (Lipton, 2004). Applied to psychoanalysis: knowledge of Freud's theories perfectly *justifies* psychoanalytic redescriptions of phenomena that figure in the X-position, but it doesn't follow that the 'facts' Freud uncovered also *explain* the X-phenomenon. However justified those redescriptions may be in the light of a psychoanalytic background theory, they lack explanatory force. Confusing one's justification for the belief that q *is the case*, with an explanation of why q is the case, explains why a well-conducted psychoanalytic interpretation leaves one with the impression that nothing *else* remains to be said—that one now fully *understands* the phenomena.

'But doesn't your analysis amount to a form of social constructivism?' We strongly reject this interpretation for basically Searlean reasons. Our reconstruction assumes a firm distinction between natural facts and institutional facts, and questions global social constructivism. For starters, a Searlean analysis of the ontology and epistemology of institutional facts is part of an analysis of the ontology of social reality, and differs from the social constructivist's implausible anti-realist credo that all (scientific) facts are socially constructed. Searle assumes a firm and plausible distinction between brute facts, which exist independently of human intentionality, and institutional facts, which come into existence when human beings collectively award what Searle calls *status functions* (the referents of the terms that occupy the Y-position in declaratives) to parts of reality. A nuanced theory of social facts and social reality starts "with the fact that we're biological beasts" (Searle) and then (and only then) asks "how is it possible in a world consisting entirely of brute facts, of physical particles and fields of force, to have consciousness, intentionality, money, property, marriage, and so on" (Smith, 2001, p. 22). Social constructivism, on the other hand, is a controversial theory on the production of scientific knowledge in general, which is substantiated by expansive ontological and epistemological claims that are widely disputed (cf. Kukla, 2000; Boghossian, 2006). In (Boudry and Buekens, 2011), we defend that social constructivism is inadequate as an account of *bona fide* epistemic practices. We do, of course, acknowledge that many psychoanalysts have turned to social constructivism to re-describe and support both the practice and the status of the supporting theory (cf. Moore, 1999, for a critique of such proposals). Following Murphy (2006), we also reject that natural psychic phenomena that acquire a specific institutional status within the psychoanalytic practice are *themselves* social constructions. Conversely, our account does not *entail* that psychic phenomena themselves—the phenomena that figure in the X-position—are social constructions. First, notice that the original psychic phenomena Freud studied—from innocent dreams and slips of the tongue, to depressions and recurrent fears and delusions—are real phenomena, observable outside psychiatric and psychoanalytic contexts. The psychological phenomena that form the objects of study in psychology and psychiatry exist independently of theories about these phenomena. They may be recognized as deviant phenomena by friends and family of sufferers, though perhaps not in any manner that can be precisely described by them, and certainly not fully explicable by them. They do not come into existence because a designated group accepts their existence. Our folk psychology, which is so effective when it comes to describing the normal mind, is clearly not suited or equipped to interpret, let alone explain, the nature of and causal mechanisms underlying mental disorders and deviant behaviour.[18] Neither can social constructivism explain these facts, and neither can it explain why disorderly thoughts or behaviour are noticed in every culture (Roth and Kroll, 1986) and are, in this sense, perfectly detectable by us (again: 'detectable' does not amount to 'being adequately explained').[19] It is precisely because the proposed non-conservative extension of folk psychology proposed by psychoanalysis introduces a complex system of *institutional* facts that psychoanalysis is *not* an extension of our natural mental economy. And there is, of course, the fascinating question as to *which* version of psychoanalysis would offer the best extension of our folk psychology—a question never satisfyingly answered by proponents of the 'folk-psychological extension'-thesis.

5 Concluding remarks: self-validating thought systems

Why is a critique of psychoanalysis as a pseudo-scientific endeavour (Popper), or potentially falsifiable and falsified (Grünbaum) not enough? The claim that (Freudian) psychoanalysis contains many false, unfounded or unfalsifiable beliefs is largely taken for granted by us. Our starting point was the fascinating but largely neglected hermeneutic power of psychoanalysis, its capacity to 'understand' or 'make sense' of almost any cultural or anthropological phenomenon, a curious and problematic feature neglected by critics who confine themselves to assessing Freud's own empirical and sci-

---

[18]This explains why the proposal to view psychoanalytic concepts and the technique as a non-conservative extension of our folk psychology, which is aimed at describing and explaining irrational or pathological behaviours and delusions, proved to be so attractive. Under this description, the institutional facts Freud created fill the gap in our naturally-evolved capacity to understand ourselves and others. Explanations appealing to witchcraft or possession by the devil once filled exactly the same gap and were also met with much enthusiasm.

[19]Social constructivists also owe us an explanation for why they think that what 'contingently exists' never disappears completely, even long after the concepts and beliefs that 'created it' have disappeared from our mental economy.

entific ambitions. Belief-systems can be self-validating in many ways. A belief-system may encourage its supporters to discount contrary evidence, may have self-fulfilling effects (like placebo effects) or may partly constitute its own truth via Searlean acceptance mechanisms (a possibility explored and defended here). Each of these phenomena can be found in psychoanalysis: immunisation strategies ('you repress the Freudian insights') illustrate the first way, while the placebogenic effects of the therapy—predicted effects occur because of psychological mechanisms that involve belief in the existence of healing powers—illustrate the second self-validating ingredient. We tried to make plausible how a third self-validating mechanism is at work in psychoanalysis: we take a number of facts and phenomena proper to psychoanalysis to be indicative of how the Freudian enterprise can be seen as creating a system of unintended institutional facts, an option explicitly left open in Searle's theory of institutional facts.

Three key observations must be added to the classic 'pseudoscience' or 'falsified theory'-objection in order to illuminate the plausibility of yet a 'third way' in which psychoanalysis has self-validating aspects: first, there is a remarkable social structure that surrounds key psychoanalytic claims—how they were *introduced*, *defended*, and *shielded off* against rival psychoanalytic claims. This feature links the unscientific and mostly false character of the theory (under the 'system of natural facts'-description, i.e., presented as empirically testable facts about an independent reality) with the distinctive recognition- and acceptance-conditions proper to Searlean institutional facts. So, while there is an important sense in which a critique of psychoanalysis could stop at the point where the core beliefs are seen as either blatantly false, too vague, or unfalsifiable, an important self-validating dimension of the theory would be missed if we did not look further: the relevance of social acceptance mechanisms needed to stabilize what counts as true in psychoanalysis.

The second element is the empirical observation—illustrated in the paper—that its key concepts can be applied everywhere, that every anthropological phenomenon X can be given at least one psychoanalytic interpretation Y. This suggests that the X- and Y-components in the relevant identification are more or less arbitrary (or perhaps better: stereotypically) related, which is yet another feature of Searlean institutional facts (compare this with the fact that, within reasonable bounds, almost every physical item can count as money or as indicating a border). The Y-concepts that figure in psychoanalytic identifications *prima facie* function like theoretical concepts (because they do not refer to observable features), but they could also be taken as experience-distant concepts (in the sense defined by Geertz, 1983) with application conditions based on what is accepted as true within a community, rather than based on empirical data. This is a key reason why a 'Searlean approach' to theoretical concepts and entities in (say) high energy physics would be completely off the mark: no physicist would claim that the Higgs particle came into existence because every informed physicist believes that it exists, and the reason is that it clearly designates a theoretical concept for which we know what *would* be empirical *evidence* for its existence. Moreover, such concepts (and what they designate) are introduced in the context of highly empirical theories that are clearly falsifiable in the sense in which psychoanalysis clearly wasn't. Thirdly, if the Higgs particle were an *intended* Searlean institutional fact, it would be incoherent to seek to confirm its existence on the basis of *physical* evidence, for an institutional property Y cannot be derived from or be reduced to physical properties (let alone physical properties of its bearer X).

A key advantage of our Searlean approach to psychoanalytic facts is that it explains how and why we—qua outsiders, not qua believers—can have true or false beliefs *about a reality unintentionally created by Freud, and why those beliefs can be true or false. Although Freud alone could have developed a falsifiable theory, it required a tightly controlled community of acceptors of his claims to turn his false theories into a pseudohermeneutics that created unintended institutional facts.* A carefully steered process of acceptance (no dissidence allowed, expulsions, ...) was necessary (though of course not sufficient) for the creation and maintenance of institutional facts. Those who accepted the Freudian claims may of course not have been aware of their institutional character. On the contrary: if pressed, they will defend that the facts are natural facts. A comparison with *religious* artefacts and their properties may be helpful here: that fact that Mount X is sacred is, on a plausible account of sacredness, an institutional fact (the connection between Mount X and its sacredness is arbitrary, and X would not have been holy had there not been a community which recognizes and/or accepts that X is sacred). But those responsible for X's sacred status will, when pressed on the issue, *deny* that its sacred character is an institutional property. In fact, the very suggestion that X's sacredness is *merely* institutional might be perceived as a diffamatory remark. The advantage of taking the relevant Freudian statements as descriptions of unintended institutional facts also allows us to give charitable interpretations of Freudian claims: they enjoy true beliefs, but what *makes* their beliefs true (in the sense explained in this paper) depends on those beliefs, contrary to what *they* think that makes those beliefs true.

This might explain a potential misunderstanding about our use of 'truth' in the paper:[20] the proposition *that dreams are wish fulfilments* is, considered as an *empirical* claim about a *natural* fact, very probably false (although one never knows what future neurosciences will teach us!), but

---

[20]Thanks to an anonymous referee for pressing us on this issue.

the proposition *that dreams count as wish fulfilments* is true in a Freudian context, and that is what we would expect if the statement in which it figures is taken to be a speech act with declarative force that creates wish-fulfilling dreams when duly accepted by others (Searle, 1995, 2010). When accepted/recognized by a community, dreams become wish fulfilments, just as pieces of paper become money under specific acceptance conditions. Put in a nutshell: without joint acceptance, the classic empiricist critique of Freud remains valid. With joint acceptance in view, a new and very powerful self-validating belief system comes in view, one that connects the arbitrary connection between X and psychoanalytic phenomenon Y, the distinctive social culture of psychoanalysis and its capacity to 'understand' everything. And note that, just as money has a function, psychoanalytic interpretations have functions: they help you 'understand' phenomenon X, allow you to see X in a different light, learn you 'how to live with X', they create permissions and obligations proper to becoming 'a psychoanalytic patiens', etc. (Searle, 2010).

As one referee suggested, shamanism is, like psychoanalysis, a belief system that is likely to collapse if people learn there are no spirits and the perceived effects are simply placebo effects, and insofar as psychoanalysis is like shamanism, its content is not about institutional facts, although the roles of a shaman or of a psychoanalyst are institutionally defined. Our approach suggests an empirical hypothesis to the effect that if a practice is exposed as clustering around unintended institutional facts—and part of exposing it may consist in debunking the explicit scientific aspirations of the theory—the practice might gradually disappear or collapse. But that need not always be the case, and there are more options. One reaction to the 'constructive' character of psychoanalytic interpretations was to appeal to global social constructivism: every scientific theory (according to this immunisation strategy) constructs its own reality, hence the 'institutional facts' created by Freud and his followers are in no sense different from 'scientific facts' (cf. Moore, 1999, for criticism of this move from within psychoanalysis). Our approach partly explains why so many psychoanalysts were quite happy to redescribe their project in broadly social constructivist terms when Popper and Grünbaum exposed the underlying theory as pseudo-science or largely falsified. Their cognitive strategies are comparable with those of religious believers who, having been convinced that their core beliefs were false or unfounded, saw an welcome opening in 'symbolic' readings of its central claims. We are inclined to see this as further empirical evidence for the claim that when a system of beliefs does not correspond with an independent reality, *it need not necessarily collapse*: the truth-makers of the central claims (abstract institutional facts) can come to be seen as created and maintained by collective acceptance of the relevant beliefs. Our diagnosis leaves open the possibility that psychoanalytic claims can be accepted as a system of *intended* institutional facts, just as religious believers can continue to accept that a mountain is sacred despite the fact that they have come to see its holiness as grounded in social rather than metaphysical facts. But we don't think this can last for long.

Bibliography

Anscombe, G. E. M. (1958). On brute facts. *Analysis*, 18:69–72.

Barrett, C., editor (1967). *Ludwig Wittgenstein. Lectures and conversations on aesthetics, psychology and religious belief.* University of California Press, Berkeley, CA.

Boghossian, P. (2006). *Fear of knowledge.* Oxford University Press, Oxford.

Borch-Jacobsen, M. (2005). Une théorie zéro. In Meyer, C., editor, *Le livre noir de la psychanalyse*, pages 178–183, Paris. Les arènes.

Borch-Jacobsen, M. and Shamdasani, S. (2006). *Le dossier Freud.* Les empêcheurs de penser en rond, Paris.

Boudry, M. and Buekens, F. (2011). The epistemic predicament of a pseudoscience: Social constructivism confronts Freudian psychoanalysis? *Theoria*, 77(2):159–179.

Breger, L. (2000). *Freud. Darkness in the midst of vision.* John Wiley & Sons, New York.

Carruthers, P. and Smith, P., editors (1996). *Theories of theories of mind.* Cambridge University Press, Cambridge.

Cioffi, F. (1998). *Freud and the question of pseudoscience.* Open Court, Chicago, IL.

Crews, F. (2006). *Follies of the wise. Dissenting essays.* Shoemaker & Hoard, Emeryville, CA.

Dupré, J. (2004). The miracle of monism. In Caro, M. D. and Macarthur, D., editors, *Naturalism in Question*, pages 36–59, Cambridge, MA. Harvard University Press.

Esterson, A. (1993). *Seductive mirage.* Open Court, Chicago, IL.

Flieger, J. A. (2005). *Is Oedipus online? Sigmund Freud after Freud.* MIT Press, Cambridge, MA.

Freud, S. (1933). *New introductory lectures on psycho-analysis*. L. and Virginia Woolf at the Hogarth Press, and the Institute of Psycho–analysis, London.

Geertz, C. (1983). *Local knowledge. Further essays in interpretive anthropology*. Basic Books, New York.

Haslanger, S. (2006). What good are our intuitions? Philosophical analysis and social kind. *Aristotelian Society Supplementary Volume*, 80:89–118.

Hempel, C. (1965). *Studies in the logic of explanation*. Free Press, New York.

Humberstone, L. (1992). Direction of fit. *Mind*, 101:59–83.

Hutto, D. and Ratcliffe, M. (2007). *Folk psychology re-assessed*. Springer, Dordrecht.

Jopling, D. (2008). *Talking cures and placebo effects*. Oxford University Press, Oxford.

Kukla, A. (2000). *Social constructivism and the philosophy of science*. Routledge, London.

Kusch, M. (1999). *Psychological knowledge: A social history and philosophy*. Routledge, London.

Lagerspetz, E. (1989). *A conventionalist theory of institutions*, volume 44 of *Acta Philosophica Fennica*. Societas Philosophica Fennica, Helsinki.

Lagerspetz, E. (2006). Institutional facts, performativity and false beliefs. *Cognitive Systems Research*, 7:298–306.

Lipton, P. (2004). *Inference to the best explanation*. Routledge, London. 2nd ed.

Macmillan, M. (1997). *Freud evaluated. The completed arc*. MIT Press, Cambridge, MA.

Masson, J. M. (1985). *The complete letters of Sigmund Freud to Wilhelm Fliess 1887–1904*. Harvard University Press, Cambridge, MA.

Moore, R. (1999). *The creation of reality in psychoanalysis*. Analytic Press, Hillsdale, NJ.

Murphy, D. (2006). *Psychiatry in the scientific image*. MIT Press, Cambridge, MA.

Olsen, S. H. (1986). *The structure of literary understanding*. Cambridge University Press, Cambridge.

Pettit, P. (1993). *The common mind. An essay on psychology, society and politics*. Oxford University Press, Oxford.

Roazen, P. (1975). *Freud and his followers*. Knopf, New York.

Roth, M. and Kroll, J. (1986). *The reality of mental illness*. Cambridge University Press, Cambridge.

Searle, J. R. (1969). *Speech acts*. Cambridge University Press, Cambridge.

Searle, J. R. (1995). *The construction of social reality*. Free Press, New York.

Searle, J. R. (2001). *Rationality in action*. Cambridge University Press, Cambridge.

Searle, J. R. (2010). *Making the social world: The structure of human civilisation*. Oxford University Press, Oxford.

Smith, B. (2001). Fiat objects. *Topoi*, 20:131–148.

Smith, B. (2003). John Searle: From speech acts to social reality. In Smith, B., editor, *John Searle*, pages pp. 1–33, Cambridge. Cambridge University Press.

Stern, D. B. (1992). Commentary on constructivism in clinical psychoanalysis. *Psychoanalytic Dialogues*, 2:331–363.

Strachey, J., editor (1959). *The standard edition of the complete psychological works of Sigmund Freud*. Hogarth Press, London.

Stroud, B. (2004). The charm of naturalism. In Caro, M. D. and Macarthur, D., editors, *Naturalism in question*, pages 21–36, Cambridge, MA. Harvard University Press.

Tuomela, R. (2002). *The philosophy of social practices. A collective acceptance view*. Cambridge University Press, Cambridge.

Winter, S. (1999). *Freud and the institution of psychoanalytic knowledge*. Stanford University Press, Stanford.

FotFS VII

François, K., Löwe, B., Müller, T., Van Kerkhove, B., editors,
Foundations of the Formal Sciences VII
Bringing together Philosophy and Sociology of Science

Looking for Busy Beavers. A socio-philosophical study of a computer-assisted proof

LIESBETH DE MOL[*]

Centrum voor Wetenschapsgeschiedenis, Universiteit Gent, Blandijnberg 2, 9000 Gent, Belgium
E-mail: elizabeth.demol@ugent.be

"Young man, in mathematics you don't understand things, you just get used to them"
John von Neumann

1 Introduction

What exactly is the impact of the computer on mathematics? If one were to believe some of the advocates of computer-assisted mathematics "computers [are] changing the way we do mathematics" (Borwein, 2008). This alleged change concerns a shift in perspective on mathematical knowledge and the way it is attained, a change brought about and made explicit through the computer. Mathematicians like Borwein and Bailey (2003, 2004), Seiden (1998) and Zeilberger (1993) have emphasized on several occasions that the increasing significance of computer-assisted mathematics makes it more and more clear that quasi-empirical or experimental methods must be included and be taken more seriously within the mathematical discourse, that mathematics has more in common with the empirical sciences than is usually believed. They question the traditional ideas on mathematical certainty, proofs, rigor and understanding. Also, within the philosophy of mathematics, (examples of) computer-assisted mathematics (are) is mainly discussed in the context of work that can be placed in the tradition of Pólya and Lakatos,[1] work that emphasizes the significance of the practice

---

[*]This research was supported by the *Fonds Wetenschappelijk Onderzoek Vlaanderen*, Belgium.
[1]Cf., e.g., Lakatos (1976) and Pólya (1954).

Received by the editors: 14 February 2009; 20 September 2009.
Accepted for publication: 21 September 2009.

of the mathematician,[2] the fallibility of mathematics and the significance of (quasi-)heuristics opposing the more traditional idea that mathematics is without history, without change, that mathematics is no more than a body of absolute and certain knowledge.[3]

The idea that computer-assisted mathematics has this kind of impact on mathematics and its philosophy has of course also been opposed. Indeed, several philosophers and mathematicians have argued for several different reasons that the computer does not have this kind of epistemological impact on mathematical methods. In the end, theoretically, it is not capable of anything we were not already capable of before.[4]

The aim of this paper is to contribute to the question concerning the impact of the computer on mathematics, a question which, in our opinion, will only gain in importance in the future. Here, I will focus on computer-*assisted* proofs. This work is to be situated in a larger research project that aims to develop a more systematic and complete approach towards computer-assisted or, as we shall identify it in the remainder of this paper, mechanized mathematics.[5] Such approach is still lacking in the literature.

One important part of such a systematic approach towards mechanized mathematics is the micro-analysis of (well-known and less well-known) examples throughout the history of mechanized mathematics, starting from (but not necessarily ending with) the accounts of the mathematicians themselves. There are already some examples in the literature of relatively detailed case studies like MacKenzie's socio-history of the four-color theorem (MacKenzie, 1999), probably the most famous example of a computer-assisted proof, and Van Bendegem's account of the Collatz problem (Van Bendegem, 2005), which is a typical example of a problem studied with the help of the computer. In this paper we shall look at a relatively unknown example of a *computer-assisted proof*, i.e., the solution of the Busy Beaver problem for the class of Turing machines with 2 symbols and 3 and 4 states. The Busy Beaver game (or competition) for a certain class of Turing machines (with m states and n symbols) is to find the Turing machine which prints out the maximum number of 1s before halting when started from a blank tape (cf. §2.1 for the technical details).

As is stated in the title of this paper, the case analysis should be regarded as a socio-philosophical analysis. This means here that I will start from a relatively detailed (micro)-analysis of a specific example of a computer-

---

[2]Cf., e.g., Van Kerkhove and Van Bendegem (2008).

[3]Cf., e.g., Tymoczko (1979).

[4]Cf., e.g., Baker (2008); Burge (1998); Detlefsen and Luker (1980); Levin (1981); Swart (1980); Teller (1980). Note that this does not necessarily mean that they oppose the idea of mathematics being not that absolute body of truths.

[5]I follow Derrick Henry Lehmer (1966) here, a computer pioneer and number theorist, who is one of the main inspirators of the present work.

assisted proof, tracing the immediate consequences and problems as they are interpreted by the mathematician(s) her/himself during the process of proving and in the communication of the proof. I will then discuss the most important philosophical problems related to computer-assisted proofs in the context of the case analysis.

1.1 What are computer-assisted proofs?

Within the literature one can easily determine different kinds of computer 'proofs'. There is, for example, an important difference between a (1) mechanized probabilistic proof that shows that a certain very large number x is prime, using the Miller-Rabin primality test, (2) visual proofs as they for example occur in fractal geometry (cf., e.g., De Mol, 2005), (3) McCune's proof of the Robbins algebra conjecture which relies on an automated theorem prover (McCune, 1997) and (4) Hales' proof of the sphere packing problem (Hales, 2005).

In this paper, unless indicated otherwise, we shall use the term *computer-assisted proof* in the sense of Lehmer (cf., e.g., Lehmer, 1963), who was involved with one of the first true computer-assisted proofs (Lehmer et al., 1962). A computer-assisted proof is a proof that proves a theorem that *practically* could not have been (or, thus far, has not been) proven, re-proven or verified by human mathematical reasoning alone. I.e., (certain parts of) both the process that results in the proof as well as the proof itself must be *humanly impractical*. As a consequence, these proofs are, practically speaking, not surveyable by humans. Furthermore, the proof is also *machine impractical* in that, besides the programming, certain parts of the proof could not have been done by the computer. In this sense, we use the term computer-*assisted* proof rather than computer proof. These proofs typically involve the verification of a large number of cases by the computer, although the work of the computer is not restricted to this verification. A well-known (and probably the most famous) example is the proof of the four-color theorem (4CT for short) by Appel and Haken (Appel and Haken, 1977; Appel et al., 1977).[6]

This definition is not intended as a once-and-for-all-given definition. It should be understood as an instrument to evaluate and demarcate certain computer applications which, when analyzed, can in their turn change the semantic content of computer-assisted proofs.

---

[6] Note that several examples from the literature that are quite frequently considered as examples of computer proofs are excluded by this definition. For example, if a computer finds a counter-example to a certain conjecture, this does not count as an example of what is here understood under computer-assisted proof as, once the counterexample has been found, the human can easily check that it is a counterexample, and thus disprove the conjecture.

1.2 Why busy beavers?

There are two main reasons that motivate my choice for this specific case analysis.

The first and least important one is that in the present case study the proofs and the problem itself can be relatively easily explained. Unlike, for example, the computer-assisted proof of the sphere packing problem, the 'proofs' are simple enough to be explained up to a relatively high level of detail. The case-study is thus ideal for the intended micro-analysis.

The main reason for selecting this case is that it is not well known. In the context of computer-assisted proofs, and, more generally, mechanized mathematics one focuses mainly on the more famous examples and neglects the more 'normal' ones. Within the literature on computer-assisted proofs, discussions are usually restricted to one of the following three examples, in order of increasing popularity:

(a) The non-existence of a finite projective plane of order 10 by Lam et al. (1989).

(b) The sphere packing problem by Thomas Hales (2005).

(c) The 4CT by Appel and Haken (Appel and Haken, 1977; Appel et al., 1977) and its alternative proofs by Robertson et al. (1997) and Gonthier (2004).

(a) is mostly only mentioned without any real discussion, it is yet another example of a computer-assisted proof, while (b) and especially (c) have given rise to several different heated debates, going from the question whether such proofs are really proofs to the problem of the refereeing of such proofs.

Of course, one cannot deny that, e.g., the 4CT is more interesting than the Busy Beaver example to be studied here since, on the one hand, the result is a proof of an old mathematical problem with a long history, and, on the other hand, it is not situated within the context of theoretical computer science (as is the case for the present case study). The same goes for the other two. However, even if (a), (b) and (c) have already been studied in the literature and are more interesting, this is no reason not to be interested in less well-known and interesting examples of computer-assisted proofs. First of all and generally speaking, if one restricts the attention to only the 'famous' examples, this might lead to an all too restrictive view on computer-assisted proofs. Secondly, by drawing the attention to the fact that there are more than three computer-assisted proofs, one shows that computer-assisted proofs are not as abnormal as one might believe and one thus counters the argument that, as there are only a few computer-assisted proofs, they cannot be that important. And abnormal they are not. A very simple search on the following terms: "Computer proof", "Machine proof",

"Mechanical proof", "Computer-assisted proof", and "Automated proof" in MathSciNet and DBLP, two on-line databases,[7] resulted in Table 1:

	DBLP	MathSciNet
"Computer proof"	196	73
"Machine proof"	42	16
"Mechanical proof"	24	53
"Computer-assisted proof"	22	161
"Automated Proof"	72	79
Total	356	382

TABLE 1. Overview of the number of papers in MathSciNet and DBLP that mention computer proofs.

Although these numbers are not overwhelming, and it is furthermore not the case that all of these proofs are computer-assisted proofs in the sense used here, they show that one cannot simply discard the significance of computer-assisted proofs on the basis of numbers. Although they have not yet become a 'normal' method of mathematics, they are more important than is usually believed.

Thirdly, by focusing on the less well-known examples, it becomes possible to study the impact of the computer not only on the level of the great innovations of mathematics—the famous examples—but also on the level of the 'everyday' practice of the mathematician, which is not redundant *if* one accepts the view that mathematics cannot be reduced to its great achievements.

Finally, it is not the case that if one has seen one very important computer-assisted proof, one has seen them all. By studying more examples of computer-assisted proofs it becomes possible to tackle certain more general questions more exactly. What kind of problems can be solved with computer-assisted proofs? What kind of methods are used and in what way are they different from other methods? How are computer-assisted proofs communicated and perceived? What kind of techniques are used to convince the fellow mathematicians of the proof? Is there an evolution in the way computer-assisted proofs are made and formalized? Etc. These kinds of questions allow to get a more concrete view on what computer-assisted proofs are and in what sense they really differ from other proofs. General questions like these cannot be answered properly if one restricts ones attention to only three computer-assisted proofs.

---

[7]MathSciNet is a database for mathematics in general, whereas DBLP is a computer science database.

2 Looking for busy beavers

In a paper titled *On non-computable functions*, Tibor Radó (1962) proposed an example of "the phenomenon of non-computability in its simplest form", an example which is now known as the Busy Beaver function $\Sigma(m)$ (for $m \in \mathbb{N}$). He provided the following motivation for formulating and studying the problem (Radó, 1963):

> Let us note that our main objective is to observe the phenomenon of non-computability in its simplest form, so that we can use the insight we achieve to see better what tasks we can delegate to computers. Actually, the comments to be presented here originated with the writer's studies relating to the optimal design of automatic systems, and specifically with efforts to use computers to the limit of their capabilities for this purpose.

In other words, the computer not only plays a fundamental role in the solution of specific cases of the problem, but also led to the formulation of the problem. Furthermore, as an example of an uncomputable problem, it is situated in the context of the theory of computing and is thus, on the theoretical level, closely related to the computer.

Recall that, given Turing's thesis or any other logically equivalent thesis, a problem is considered non-computable (or recursively unsolvable) iff. there is no Turing machine that is able to compute it. One of the more famous examples is the halting problem, i.e., the problem to decide (compute) for any Turing machine whether or not that machine will halt.

The fact that a problem is non-computable in general, does not mean that every instance of the problem is also non-computable. I.e., it is not because there is a Turing machine with a non-computable halting problem that every Turing machine has a non-computable halting problem. One can thus search for strategies that allow to decide a certain generally undecidable problem for specific classes of 'decidable' Turing machines. This is the goal Radó and, after him, several other researchers, set themselves: to compute the Busy Beaver function $\Sigma(m)$ for specific numbers m. It was soon understood that the computer would be an indispensable helper.

After a preliminary section (§2.1), defining some of the basic notions used here, we shall give a (relatively) detailed account of computer-assisted proofs that $\Sigma(3) = 6$ and $\Sigma(4) = 13$ (§2.2). This will be followed by a discussion of three of the typical features of these proofs (§2.3). In the next section (§3), we shall confront these proofs with some of the fundamental epistemological problems related to computer-assisted proofs.

2.1 Some preliminaries

A (standard) Turing machine T consists of a read-write head and a two-way infinite tape. A blank is denoted by the symbol 0. To start with, the tape contains a finite initial configuration, possibly empty, on an otherwise blank tape. In its initial state (state 1), the head reads the leftmost symbol of the initial configuration.[8] The machine is said to halt when it reaches the halting state H. T is formally defined by the finite set of states Q plus the halting state H, a finite set of symbols $\Sigma = 0, 1, ...$ and a transition function $f : Q \times \Sigma \to (\Sigma \times \{L, R\} \times Q)$. The transition function f determines for any state $q_i \in Q$ and any symbol $s_j \in \Sigma$ what the machine should do when in state q_i, reading the symbol s_j. I.e., if $f(q_i, s_j) = (s_{i,j}, D_{i,j}, q_{i,j})$ then, if T is in state q_i, reading symbol s_j, T replaces s_j by $s_{i,j}$, moves in direction $D_{i,j} \in \{L, R\}$ (L stands for left, R stands for right) and goes to state $q_{i,j}$. In what follows, an instruction of a Turing machine will be represented by the quintuple $(q_i, s_i, s_{i,j}, D, q_{i,j})$.

m	$S(m)$	$\Sigma(m)$	Source
1	1	1	Trivial
2	4	2	Mentioned by Radó (1962)
3	21	6	Lin and Radó (1965), Brady (1983), Kopp (1981)
4	107	13	Brady (1983), Kopp (1981)
5	$\geq 47\,176\,870$	≥ 4098	Marxen and Buntrock (1990)
6	$> 3.8 \times 10^{21132}$	$> 3.1 \times 10^{10566}$	unpublished (May 2010)

TABLE 2. Overview of the current result in the Busy Beaver competition. The 2010 bound is attributed on Pascal Michel's webpage on Busy Beavers to Kropitz.

Let $\mathrm{HT}(m, 2)$ be the class of Turing machines with m states and 2 symbols that halt when started from a blank tape. Then, for $T \in \mathrm{HT}(m, 2)$ let $\sigma(T)$ and $s(T)$ denote the number of symbols different from 0 left on the tape and the number of computation steps before T halts, respectively. Let $\Sigma(m)$ be the maximum $\sigma(T)$ and $S(m)$ the maximum $s(T)$ with $T \in \mathrm{HT}(m, 2)$.

Definition 1. The Busy Beaver problem is the problem to determine $\Sigma(m)$ for any $m \in \mathbb{N}$

Definition 2. The maximum shift number problem is the problem to determine $S(m)$ for any $m \in \mathbb{N}$

Both problems were proven to be uncomputable by Radó (1962).[9] Note that computing specific values $\Sigma(m)$ and $S(m)$ for specific n comes down to

---
[8]Of course, if the initial configuration is empty, the head starts at some arbitrary square, reading 0.
[9]Radó only considered 2-symbolic Turing machines. The reason for this is that any n-symbolic Turing machine can be simulated or reduced to a 2-symbolic Turing machine.

solving a special case of the halting problem for the class of machines with n states as one needs to be able to determine the subclass $HT(m, 2)$. Table 2 gives an overview of the known values in the Busy Beaver competition.

2.2 Determining $\Sigma(m)$

> [...] when the writer wanted to find a certain highway on an automobile trip, he received the following directions [...]: "Drive straight ahead on this road; you will cross some steel bridges; and after you cross the last steel bridge, make a left turn at the next intersection." Luckily, the unsolvable problem implied by this advice was resolved by a member of the construction crew who volunteered the information that "after you cross the last steel bridge, there isn't another steel bridge until you reach Richmond, 130 miles away." (Radó, 1962)

In his paper (Radó, 1962), Radó pointed out that the case with $m = 1$ is trivial and that the case $m = 2$ was computed during a seminar. For any $\Sigma(m), S(m)$ with $m > 2$, laborious and lengthy proofs, including long computations, seemed unavoidable. The reason for this is that the number of Turing machines with 2 symbols and m states grows exponentially fast for increasing m. Indeed, the size of the class of 2-symbol Turing machines with m states is equal to $(4m + 1)^{2m}$.

Lin and Radó (1965) proved with the help of the computer that $\Sigma(3) = 6, S(3) = 21$. Brady (1983) and Kopp (1981) (reported in Machlin and Stout, 1990)[10] proved that $\Sigma(4) = 13, S(4) = 107$ and also confirmed the results by Radó and Lin. In what follows, we shall give a relatively detailed account of the proofs of these results.

Before doing so, I must point at a difference between the Turing machine representation used by, on the one hand, Brady and, on the other, Kopp, Radó and Lin. Contrary to Brady, the latter treat a halt as a separate branch to a state 0 within a normal entry of a Turing machine. This has an effect on the total number of 2-symolic Turing machines with m states. Instead of $(4m+1)^{2m}$, there are now $[4(m+1)]^{2m}$ distinct 2-symbol, m-state machines. Table 3 gives an overview of the total number of machines with 2, 3 and 4 states for both approaches.

As is clear from Table 3 the approach by Brady results in a smaller number of cases to start from.

The three proofs all make use of a series of computer-assisted reductions of the total number of cases for each of the classes of 2-symbolic Turing machines with 3 and 4 states, until finally no so-called *holdouts* remain.

This was proven by Shannon. In current research on the Busy Beaver problem, one also considers Busy Beaver functions for classes of Turing machines with the number of symbols $m > 2$; cf., e.g., Michel (2004). We shall not consider these generalized Busy Beaver problems here.

[10]"Kopp" is the maiden name of Machlin.

m	$(4m+1)^{2m}$	$[4(m+1)]^{2m}$
2	6561	20736
3	4 826809	16 777 216
4	6 975 757 441	25 600 000 000

TABLE 3. Number of 2-symbol Turing machines with m states.

I.e., it was determined for each of the machines individually, whether or not they halt when started from a blank tape, and, if they halt, which values $s(T)$ and $\sigma(T)$ they have. For each of the proofs found, there was a conjectured value for $\Sigma(m)$ and $S(m)$. These were proven lower bounds, found by making use of certain heuristic and explorative methods.[11] In all cases, the conjectured values turned out to be the correct values.

The proofs consist of two main stages of reduction, a more theoretical and a more heuristic stage. In the first stage, certain theoretical considerations are used that result in the immediate (computer-assisted) elimination of a large number of machines. Radó and Lin came to the conclusion that all machines the first instruction of which is not $(1, 0, 1, R, 2)$ could be eliminated.[12] These methods were also used by Brady and Kopp. They extended the argument by Radó and Lin by what they call a *tree generation*, generating instructions as they are needed.[13] This method could be easily automated and was used to eliminate a large number of machines. Table 4 gives an overview of the number of remaining machines, called the *holdouts*, after the application of the several elimination methods used in stage one for each of the proofs.[14] The differences between the number of remaining machines after application of the tree normalization between Brady and Kopp can be explained by slight differences in their respective approaches. As is clear from Table 4 certain theoretical considerations allowed for a serious reduction in the number of cases to be considered. However, the number of remaining cases is still too large to be humanly manageable.

The next step in all the proofs is to turn to, what Brady calls, more heuristic proof techniques. In the next stage, Radó and Lin first used the conjectured value $S(3) = 21$ in order to eliminate some further machines.

---

[11] E.g., Brady (1966) mentions that he used certain heuristic methods to conjecture that $\Sigma(4) = 13, S(4) = 107$. Note that in the ongoing research on the Busy Beaver competition, one still makes use of several heuristic methods to determine lower bounds, methods which are also used in proofs of Busy Beaver winners.

[12] For an explanation why this can be done the reader is referred to (Lin and Radó, 1965).

[13] For more details the reader is referred to (Machlin and Stout, 1990).

[14] The method of (tree) normalization was not the only method used. However, it is the most important one.

n	Radó and Lin	Brady	Kopp
3	82,944	± 4,000	3,936
4	\	± 550,000	603,712

TABLE 4. Number of Turing machines remaining after the first series of reductions.

Clearly, all those machines that halted before this respective number of steps was reached, could be eliminated. Kopp used the same technique, although she did not use the conjectured value for $S(4)$ but decided to run each of the remaining machines for some hundred steps $n > 107$. In case of Kopp, this technique led to the elimination of 1364 and 182,604 machines for the 3-state and the 4-state case, respectively. Brady first reduced the ± 4,000 and ± 550,000 remaining machines to 27 and 5,820 *holdouts*, respectively, by coupling the tree generation program to "a heuristic solution to the halting problem". No exact numbers are known in the case of Radó and Lin.

So how to proceed from here? In the next steps, Kopp, Brady, Lin and Radó made use of more explorative and heuristic methods. These were used in order to:

- identify or 'discover' different types of 'pattern", called *infinite loops* by Kopp, in the behavior of the holdouts, and

- automate the detection of these patterns in the holdouts.

Kopp, Brady, Lin and Radó were able to identify different types of infinite loops with the help of the computer. For each of these types of loops it can be proven that, if they occur in a given Turing machine, then that machine will never halt and thus its halting problem is decided. Now, if it could be proven for each of the holdouts that its ultimate behavior is an infinite loop, then it is proven that these holdouts will never halt and one can thus prove the result (since the values $s(T)$ and $\sigma(T)$ are known for each of the halting machines).

What Kopp, Brado, Radó and Lin basically did was first to print-out the behavior of some of the holdouts, study it and try to detect certain patterns that could then be generalized and be proven to be cases of infinite loops. Programs were then written that allowed for the automated detection of infinite loops which could then result in the elimination of machines whose ultimate behavior was one of the infinite loops found and formalized in a program. In the end, several types of infinite loops were detected. The most important ones are simple loops, Christmas trees, shadow Christmas trees and counters.[15]

---

[15] Note that not all of these types were found by Radó and Lin.

Identifying and then detecting different types of infinite loops was the hardest part of the Busy Beaver proofs. Machlin (Kopp) and Stout describe it as follows (Machlin and Stout, 1990, pp. 91–92):

> The major effort in calculating busy beaver numbers [...] lies in proving that large numbers of machines are in infinite loops [i.e., never halt]. The approach taken [by Brady, Radó and Lin and Kopp] is to examine some of these machines by hand, elicit a common behavior which insures that a machine is in an infinite loop, and then write a program which examines candidate machines and proves that some of them do indeed have that behavior. This process tends to iterate, with the researcher constantly trying to reduce the number of unclassified machines by either generalizing types of behavior earlier searched for, or by discovering new types of behavior.

Brady (1983, p. 661f.) gives the following description of the structure of the automated detection of infinite loops:

> BBFILT was used to separate heuristically the 5,820 holdouts into "Xmas Trees", "Counters", and "Unknown", while BBFXX, a modification of BBFILT separated the "Alternating Xmas trees" from the "Unknown" set.
>
> BBX2 was the Xmas Tree prover [...]. BBSHAD was a modification to handle "Trees with Shadow", BBALTX was an extension of BBX2 to handle "Alternating Xmas Trees", while BBALTX1 was a minor modification of BBALTX to handle double sweeps in which the extremum was reached on alternate sweeps only.
>
> BBC was the counter prover, while BBCM was a modification of BBC to handle "two-shot" carries and some cases of cell interdependence.
>
> More than 18 other programs were written for various housekeeping purposes, simulating and displaying machine behavior, exploring other reduction and filtering possibilities, etc. In all, at least 53 files were created and maintained for the project. Keeping track of what resembled a large scientific experiment became a major task in itself.

After the infinite loop detection program was applied, the small number of remaining hold-outs were then examined "by hand" and all eliminated as other cases of infinite loops. Hence, the results that $\Sigma(4) = 13, S(4) = 107$ (in case of Brady and Kopp) and that $\Sigma(3) = 6, \Sigma(4) = 13$ (in case of Lin and Radó, Brady and Kopp). Even this last stage was partly computer-assisted. As Brady explains:

> All of the remaining holdouts were examined by means of voluminous printouts of their histories along with some program extracted features.

2.3 Some features of the proofs

In what follows I will discuss three important features of the Busy Beaver proofs in more depth.

2.3.1 Experimental and heuristic methods

As was shown in §2.2, the second stage of each of the proofs is more involved. It is also this stage which can be called the more explorative and heuristic stage of (the process of finding) the proof. As Brady (1983, p. 647) describes it:

> In this final stage of the $k = 4$ case, one appears to move into a heuristic level of higher order where it is necessary to treat each machine as representing a distinct theorem. [...] The proof techniques, embodied in programs, are entirely heuristic, while the inductive proofs, once established by the computer, are completely rigorous and become the key to the proof of the new and original mathematical results: $\Sigma(4) = 13$ and $S(4) = 107$.

The 'heuristic' character of the second stage of the proof needs to be situated on two different levels: on the one hand, the identification of different kinds of infinite loops (simple loops, christmas trees, shadowy christmas trees, counters) and their variants, on the other, the actual detection of these loops in the class of holdouts, ultimately reducing the number of holdouts to 0.

The identification of new types of infinite loops can be considered as an experimental process in the following sense: the behavior of some (randomly selected) holdouts was printed out, then examined by hand and a new type of loop or some variant of an already known type was possibly identified. Brady also mentions that the computer was not only used to merely print out the behavior to assist in the identification of infinite loops, but also to extract certain features that might indicate an infinite loop. It was not known in advance whether the holdout studied would show some new pattern of an infinite loop. Maybe more steps would be needed in order for such a pattern to show itself or maybe it was a halting machine. It was also unknown in advance how many different types and which types one could expect. To paraphrase Lehmer, this process of identification of infinite loops is a process of exploring the universe of mathematics, assisted by the computer.

The heuristic character of the actual detection of infinite loops in the number of holdouts, concerns the use of what Brady has called *heuristic programs*. These are identified as such, because there is no guarantee that the decision made by the program is the correct one. Given, for example, one of the christmas tree detection programs, then the uncomputability of the halting problem combined with the practical fact of finite time implies that it is not guaranteed that this christmas tree detection program will detect every case (or a variant) of a christmas tree. I.e., there are cases of holdouts which might actually be cases of christmas trees, but which will not be detected as such by the detection program. It might for example be the case that the holdout is a yet undiscovered variant of a christmas tree, or that the typical behavior of a (variant of a) christmas tree, as described in the christmas tree detection program only 'shows' itself after say billions and billions of computation steps. Since one does not know in advance whether a given holdout is a case of a christmas tree, nor, if it is, *when* the typical behavior of a christmas tree will be observed, one needs to make certain choices, in order to assure that the christmas tree detection program will halt for every case. Machlin and Stout explain that if their christmas tree detection program ran too many steps without finding the desired behavior then the machine remained a holdout, even if it might in fact be a case of a christmas tree. This problem is described as follows in the case of a program called the backtracking program (Machlin and Stout, 1990):

> While backtracking can be useful, it cannot be guaranteed to always stop since otherwise it would supply a solution to the halting problem. As with all the heuristics we discuss, one must make some decision as to how long to run this technique before abandoning it.

Another example of a kind of heuristic program (used by Brady and Kopp) is the 'tentative' classification of the holdouts as cases of christmas trees or counters. This classification was made on the basis of the rate at which new tape squares were visited. On the basis of the decision 'made' by this program, either a counter or christmas tree detection program was run. Brady calls this program (BBFILT) a heuristic filter, "a heuristic technique based upon experimental observation."

It has been argued that computer-assisted proofs like the 4CT show that there are certain parts of mathematics that are (quasi-)heuristic in nature. (Tymoczko, 1979) is the most well-known paper in this respect. His main reason however for considering the 4CT heuristic in nature, is the fact of its human unsurveyability. This argument has been countered in the literature on many occasions.[16] Now, the authors of the Busy Beaver proofs very clearly do not shy away from identifying certain aspects of their proofs as experimental, explorative or heuristic in nature. However, their reasons for doing so has nothing to do with the unsurveyability of their proofs, but rather with the inherent and practical unpredictability of the different Turing machines to be considered. It is this unpredictability that forces them to explore the behavior of the different machines and to use so-called heuristic programs.

---

[16] Cf. §3 for more details.

2.3.2 The process of the proof

Normally, when a proof is published in a paper, no (conscious) mention is made of the process of finding that proof.[17] The actual proof and the process of finding that proof are considered to be strictly separated. What counts is that there is a proof of some theorem. The proof is that which needs to be communicated, what needs to be published, what remains. The dirty details of the process that resulted in the proof are considered irrelevant.

In the published accounts of the Busy Beaver proofs, on the other hand, one finds that a lot of information is given about the process of finding the proof. In a certain way, this should not come as a surprise: as is clear from the respective papers, as far as Busy Beaver proofs are concerned, the process of finding the proof and the proof itself are very much intertwined. It is during the process of reducing the number of holdouts (the process of the proof itself) that new (variants of) types of infinite loops are discovered, that computer programs need to be refined or that new programs need to be written, etc.

Even though the initial intertwinement between the process of finding the proof and the proof itself is in a certain way trivial,[18] the fact that the Busy Beaver proofs are represented and communicated in terms of their discovery processes is an important (but of course not exclusive) feature of these proofs. So why choose this strategy? Brady (1983) gives the answer:

> While not all the exploratory activities are reproducible, the runs shown [...] can be reproduced, so that by utilizing the techniques described in this paper the proof can be corroborated.

In other words, information on the process of finding the proof is provided in order for the reader-mahematician to be able to verify the proof and see whether it does not contain errors. This is a very important feature as it is not only a local strategy against the problem of hardware and software errors in computer-assisted proofs (cf. §3), but also a way to 'convince' other mathematicians of the result: even though they do not have all the details they have enough information to convince themselves of the correctness of the proof.

---

[17]Of course, to say that no traces at all can be found in published mathematical papers of the process of finding and the practice underlying a proof, is caricatural.

[18]From a certain point of view, it is indeed almost trivial to say that if one is 'in' the process of searching for a proof, and this process ultimately results in a proof, then searching for that proof is also always 'making' that proof, hence the intertwinement. However, once the proof is 'found' the proof is all that needs to be represented, the proof which might be very different from the proof 'as it was found'. In the end, it is the proof that counts not the process that resulted in the proof.

2.3.3 Man-computer interaction and the machine's responsibility

A last feature that needs to be mentioned here is the fact that the respective proofs result from a complex process of man-computer-interactions.

In the first stages of the Busy Beaver proofs the process of man-computer interaction is relatively simple. The human work is strictly separated from the computer work. First, there is the theoretical and human idea of tree generation which is translated into executable computer code. The computer then generates the reduced class of Turing machines. There is only one moment of interaction: when the human translates the theoretical ideas to the computer and asks the computer to execute them. Here, one could say that the machine's role is a very passive one. It is a mere calculation, following an order. It is not very involved, it is hardly responsible for the actual result. This is probably also one of the main reasons why the computer is hardly mentioned by Radó, Lin, Brady and Machlin and Stout in describing this first stage!

In the second stage however, which is the more heuristic stage, the interactive process is far more complicated. It is through a constant process of back-and-forth interaction, using a programming language, the display and print-outs as the means of communication, the 'common languages' (interfaces) through which man and computer communicate with each other, that the proof is finally found. During this process, new types of infinite loops are discovered, new programs are written or old ones are extended, etc. In this second stage, the computer is more actively involved in the process of (finding) the proof. Here, its contribution is also more explicitly mentioned by Radó, Lin, Brady and Machlin and Stout. Although, during this process of interaction, it is relatively clear which kind of things are done by the computer and which are done by the human, one cannot say that both sides are strictly separated from each other, they are involved with each other and it is this involvement that results in a proof.

This aspect of mechanized mathematics, the way man and computer interact with each other and the machine's involvement in this process, is mostly neglected in the literature on this topic. One of the exceptions is Derrick H. Lehmer. He used the idea of the amount of machine involvement in, and responsibility for, a result to order the different reasons why a mathematician would want to add pulse circuitry to the more usual pencil-and-paper method.[19] For Lehmer, the computer's responsibility or involvement is at its highest in case of computer-assisted proofs.[20] In fact, for Lehmer,

---

[19] Cf. Lehmer (1966, pp. 745–749) and Lehmer (1969, pp. 118–119).

[20] Another similar parameter Lehmer uses to order different usages of the computer in mathematics is the question whether the mathematician, who publishes the result that was established in some or the other way with the help of the computer, will mention the

computer-assisted proofs result from a true man-computer collaboration. He described this process as follows (Lehmer, 1966):

> We are dealing here with a man-machine cooperative. The man furnishes to the machine the best information that he has about the proposed theorem and the sort of proof that he thinks is likely to succeed. From this you will infer correctly that the actual proof is unknown to the man. In fact he doesn't know whether the theorem is false, or, if true, whether the machine can prove it. The machine is asked to carry out the logical steps of the proof, if indeed it can, in the allotted time. You will infer from this that there are a great many steps and that they cannot be carried out by hand. Usually the steps are not only numerous but are connected in some complicated combinatorial way. Here we are exploiting not only the speed of the computer but also its logical circuitry that allows it to keep track of and to modify its own complicated program to a degree well beyond human capability. Theorems of this kind are not easy to find in those drab branches of mathematics where elaborate proofs are not the rule. However, there are infinitely many such theorems in number theory alone.

As is reflected by this quote, the idea of a man-computer collaboration does not mean that the computer is assigned some kind of 'artificial intelligence', or the idea that the computer is capable to simulate or really be as intelligent as a human mathematician. On the contrary, it is made explicit by Lehmer that the computer's contribution lies in doing those things we are really bad at, while the human mathematician takes care of those things the computer is bad at. The computer is involved here because it 'thinks' differently than we humans do. It is an active partner in the process of finding a proof, however, a partner that is not human and should not be or behave like a human. In fact, it is because the computer is thinking *not* like a human mathematician that it is so good at what it does! In a way, this is the perfect collaboration: getting new results by combining different talents.[21]

---

computer or not and if yes, how much responsibility he will assign to the computer. A rather extreme example in this respect are some of the papers (co-)authored by Shalosh B. Ekhad. Cf., e.g., Ekhad (1990).

[21] This point was also made by Appel in an interview on the four-color theorem: "The computer [...] was not thinking like a mathematician [...] The computer was using [...] these bits of knowledge it had in every conceivable way, and any mathematician would say, 'No, no, no, you have got to organize yourself, you have got to do it that way', but the computer wasn't doing that. And it was more successful, because it was not like a mathematician" (quoted in MacKenzie, 1999).

3 Managing unsurveyability, mathematical understanding and errors

Since the 4CT there has been a growing number of philosophical papers on computer-assisted proofs, both by philosophers and mathematicians. The main question is whether or not computer-assisted proofs like that of the 4CT change our understanding of what a proof is. I.e., do computer-assisted proofs have any fundamental impact on the epistemology of mathematical proofs and thus, ultimately, mathematical knowledge? There are three main problems in this context.

A. The problem of unsurveyability. One of the most frequently discussed problems with respect to computer-assisted proofs (basically, the 4CT) is the problem of unsurveyability. The problem comes down to what was once called the falling-tree conundrum (Calude, 2001): Has a tree fallen if no one can hear it? Has a theorem been proved if no one can read its proof, surveying every one of its details? Besides the possibility of a lengthy theoretical part (as is the case for the 4CT and the sphere packing problem) and thousands of lines of code that need to be surveyed and reviewed in order to convince oneself of the proof, computer-assisted proofs involve millions of computations that, practically, cannot be surveyed by a human. This problem has several consequences. First of all, as we do not know all the details of the proof, as we have not 'followed' the proof in all its details, one must ask how one can still understand the proof (cf. problem **B**). Secondly, as computer-assisted proofs are unsurveyable one must ask how one can be sure that they are error-free, how one can rely on a proof for which one does not have all the details (cf. problem **C**)?

On the basis of the inherent unsurveyability of the 4CT, Tymoczko argued that this is a new kind of proof. Because of its unsurveyability, the proof of the 4CT shows that there are a posteriori mathematical truths, truths which rely on empirical evidence. For Tymoczko, the proof of the 4CT shows that mathematics is fallible and empirical. The epistemological role Tymoczko assigns to the 4CT because of its unsurveyability has been opposed in the literature. Some have argued that the unsurveyability of computer-assisted proofs does not show that mathematics is fallible and empirical (cf., e.g., Levin, 1981; Swart, 1980; Teller, 1980). One of the main arguments here is that Tymoczko places too much weight on the human factor of proof. It is not because a proof is humanly unsurveyable that it is for the computer (cf., e.g., Arkoudas and Bringsjord, 2007; Levin, 1981; Krakowski, 1980). Another argument is that surveyability is actually not an essential property of proofs (cf., e.g., Detlefsen and Luker, 1980; Teller, 1980). Others have argued that although they agree with Tymoczko that empirical evidence is used in mathematics, this is not something novel introduced by the 4CT (cf., e.g., Detlefsen and Luker, 1980). A different perspective on the unsurveyability of proof is offered by Shanker (1986): he claims that a proof is only a proof when it is humanly surveyable, hence, computer-assisted proofs like the 4CT cannot be considered as proofs.

B. The problem of mathematical understanding. This problem is very closely related to problem **A**. It basically comes down to the following question: how can we achieve mathematical understanding if part of the argument is hidden in a box? According to Bonsall (1981), the fact that we do not have all the details of the proof implies that computer-assisted proofs like the 4CT are not 'real' mathematics. He says:

> It is no better to accept without verification the word of a computer than the word of another mathematician.[...] We cannot possible achieve what I regard as the essential element of a proof—our own understanding—if part of the argument is hidden away in a box.

Halmos (1990) shares this opinion. For him, computer-assisted proofs like the 4CT are like oracles. All you know is that the 4CT is true but you do not know why it is true. He goes on

> I feel that we, humanity, learned mighty little from the proof; I am almost tempted to say that as mathematicians we learned nothing at all. Oracles are not helpful mathematical tools.

Related to this is the idea that theorems like the 4CT seem to have a certain arbitrariness, it is not clear why they are true exactly.[22]

C. The problem of (hardware and software) errors. Because of the unsurveyability of computer-assisted proofs like the 4CT (the length of the programs and computations), these computer-assisted proofs also suffer from the further defect that one cannot be sure that no errors have occurred (cf., e.g., Bonsall, 1981; Tymoczko, 1979). It is well-known that checking the correctness of a program is a very hard problem. Besides, it is known that hardware errors do occur. When very long computer programs and computations are involved, there is thus a real chance of machine and human errors (cf., e.g., Lam, 1991).

These three problems are very real and the debates arising from them are still open. As is clear from this summary, there are reasons to accept and reasons to reject the idea that computer-assisted proofs lead to a fundamental change of the notion of proof in mathematics, and thus have a fundamental impact on the foundations of mathematics.

---

[22]The idea that there are mathematical truths that are true for no clear reason, that are in a certain way random, has been advocated by Gregory Chaitin, one of the founders of algorithmic information theory. He uses the definition of randomness from algorithmic information theory in order to make his point.

Also in case of the Busy Beaver proofs these problems are real. In fact, one can say that, by definition, computer-assisted proofs will always suffer from these problems. In the end, they are humanly impractical, so some part of them must be unsurveyable by humans, and we thus also automatically get the problem of mathematical understanding. The problem of (software and hardware) errors also seems unavoidable, although there are some recent examples that avoid this problem to some extent.

Although a detailed philosophical analysis of these three problems on a macro level is necessary in order to come to terms with the impact of computer-assisted proofs on the notion of proof, it is also interesting to see how these problems pop-up in different computer-assisted proofs,[23] and, more importantly, how they are or can be dealt with locally. In the end, as these three problems are real, they need to be managed in some way or other at the more local level of finding, representing and communicating a proof. Tracing these 'smaller' changes can at least help to clarify the changes computer-assisted proofs bring about at the level of the practice itself.

So how how do these three problems occur in the context of the Busy Beaver proofs and how are they managed?[24] In a certain sense, these proofs are maybe not as unsurveyable as the original 4CT since the theoretical parts of the proofs are relatively short.[25] However, the proofs remain unsurveyable. First of all, they involve a lot of human work hardly any of the details of which are published. For example, the 210 holdouts in the Kopp proof are all proven to be cases of infinite loops by hand. Clearly, if the proofs of these 210 holdouts would have been included in the published accounts of these proofs, tens of pages would need to be added, including the print-outs of the behavior of these 210 holdouts. Secondly, the proofs involve very long (unpublished computer) code. Finally, and this is an essential feature of computer-assisted proofs, the proofs involve thousands if not millions of computations that result in thousands of different small proofs determining for each of the Turing machines whether or not they will halt. Taking these three aspects together, it is clear that Busy Beaver proofs are unsurveyable.[26] Since the problem of mathematical understanding and in-

---

[23] And not just in the one case one usually considers.

[24] I will mainly focus on the papers by Brady (1983) and Machlin (Kopp) and Stout (1990).

[25] The original proof of the 4CT involved not only millions of computations and computer code that initiated them but also a considerable amount of other 'text'. As is explained by Appel and Haken: "This leaves the reader to face 50 pages containing text and diagrams, 85 pages filled with almost 2500 additional diagrams, and 400 microfiche pages that contain further diagrams and thousands of individual verifications of claims made in the 24 lemmas in the main sections of text" (Appel and Haken, 1986).

[26] Of course, the first two features are not exclusive for nor essential to computer-assisted proofs.

sight is very closely related to the problem of unsurveyability, it also occurs in case of the Busy Beaver proofs. The same goes for the problem of human and computer errors (problem **C**). So, how are these problems managed in the context of the Busy Beaver proofs discussed in the previous section?

I. Managing the problem of error. In the paper by Machlin (Kopp) and Stout it is noted that:

> The work in Kopp (1981) is an independent confirmation of Brady's results, which is important since the sheer volume of human and computer work involved raises the possibility of error.

Brady makes a similar comment, pointing at the necessity of independent verification:

> Proofs of 'correctness' of the programs used are not practical. Independent verification is the only means we currently have at our disposal.

As is clear, Brady, Machlin (Kopp) and Stout are very much aware of the problem of (hardware, software and human) errors. Indeed, because of the length of the actual proof (the programming, the execution of the code in several stages (!) and the human work to prove the final holdouts) the chances that an error has occurred increase, and the result might thus be false. Furthermore, the unsurveyability of the proof also makes it impossible to check every detail of the proof and conclude that the proof is 100% watertight. There are several ways to deal with these problems. The best solution I know of to avoid errors is the use of interactive theorem provers like HOL to attain formal proofs that have been checked completely by the computer.[27] The 'method' mentioned by Brady, Machlin and Stout to reduce the chance that (human or machine) errors have occurred is that of independent verification. As was explained in §2.3, the description of the process of the proof can also be understood as a strategy to allow other mathematicians to check the correctness of the proof.

Although the method of independent verification can never lead to complete certainty, it is often one of the more efficient methods available. Even if one cannot exclude the possibility of errors in the Busy Beaver proofs, the fact that the cases $\Sigma(m)$ and $S(m)$ for $m < 5$ were verified, not only reduces the chances of errors but, maybe even more so, says something about the

---

[27]For example, the last version of the 4CT by Gonthier is such a proof. Also Thomas Hales has started the FlySpeck project in order to produce a completely formalized and computer-checked proof of the sphere packing problem, in a reaction to the fact that after a team of referees had worked on the proof for 15 years, they concluded that they were only 99% sure that it contains no errors. For a philosophical discussion of formal computer-checked proofs related to the problem of unsurveyability and error, cf. Arkoudas and Bringsjord (2007).

way mathematicians think they should deal with one of the typical problems of computer-assisted proofs.[28]

II. Managing unsurveyability. As is explained by Brady in the next-to-last quote of §2.2, "keeping track of what resembled a large scientific experiment became a major task in itself". For this reason, Brady wrote more than 18 programs for, among other things, "various housekeeping purposes". In other words, it is clear that the length of the whole proof was really a problem which made it necessary to implement certain strategies to deal with the complexity of the proof.

The problem of unsurveyability is also a problem if one needs to write down and communicate the results one has found. One needs to find a good format to present the proof in, despite its unsurveyability. Contrary to the famous papers by Appel and Haken, or that by Hales, the published papers on Busy Beaver proofs considered here are relatively short despite their alleged unsurveyability. One of the reasons for this is that not that many details are provided in the respective published versions of the proofs.[29] This not only concerns the computer programs and the actual computations. For example, only the main types of infinite loops are described in detail, despite the different kinds of variants discovered for each of the types. Also, the details of the proofs of the holdouts that were proven by hand are not provided. To the sceptical reader, this can only add to the mistrust one must have in this kind of proofs, making them even less suitable for review than for example Thomas Hales' proof of the sphere packing problem. However, from another point of view, one could say that providing an explicitly short 'summary' of a computer-assisted proof, skipping a lot of the details, is just another way to deal with the unsurveyability and thus the impossibility of communicating every single details of the proof. By dealing with this problem in this way, the 'persuasiveness', the argumentative power of these proofs, as they are communicated to the reader, functions differently. Even though not all the details are provided, the reader is given enough information to understand how the proof works. The papers describing the proofs provide a general descriptions of the programs, how they are used for the tree reduction, for the detection of infinite loops or the exploration of the behavior of the Turing machines studied, the order in which these programs were executed, etc. This way of communicating and writing down computer-assisted proofs, a method which is in a certain way not new to

---

[28]It should be pointed out that they are not the only ones to emphasize the significance of independent verification. E.g., Lam (1991) points out that "[the proof of the non-existence of the projective plane of order 10] is only an experimental result and it desperately needs an independent verification, or, better still, a theoretical explanation".

[29]The technical note describing Brady's proof in more details and Kopp's PhD containing the proof are of course lengthier.

mathematics,[30] is an important possibility for dealing with the inherent unsurveyability of these proofs. The unsurveyability of course remains, and the proof will not be represented in all its details. However, the problem is dealt with by developing a style, a method of representing these proofs that still allows them to be published in a reasonable amount of pages and be reviewed in a reasonable amount of time, still allowing the reader to understand the main techniques of the proof. Of course, the reviewer will need a certain amount of trust if he/she does not want to go through the trouble of reconstructing every detail of the proof.[31] He/she has to trust not only the machine, but also, and maybe even more, the mathematician.

The question of what to include and what not to include in a paper describing a computer-assisted proof is a very intricate one and needs further consideration. However, if computer-assisted proofs are ever to become part of the common discourse of mathematics, published papers of proofs that need over 100 pages will have to be the exception rather than the rule, else, the communication of mathematical knowledge will become practically impossible.

III. Managing the problem of understanding. A problem that is very closely related to the unsurveyability of computer-assisted proofs is the problem of mathematical insight and understanding, the question of the explanatory power of computer-assisted proofs. The same question must be posed in the context of the Busy Beaver proofs. Indeed, in how far does, e.g., Kopp's computer-assisted proof provide an understanding of the fact that of $\Sigma(4) = 13$? From a certain point of view, the answer to this question is that the Busy Beaver proofs do not really provide an understanding of the facts proven. Since large portions of the proof are 'generated' by the computer without any human mathematician having surveyed this part of the proof we can never fully understand why $\Sigma(4) = 13$. This problem seems to become even worse if one has not 'found' the proof but has merely read the descriptions of the proofs by Radó, Lin, Brady and Kopp. However, to conclude on this basis that these Busy Beaver proofs do not provide any insight or understanding whatsoever, that, to put it in Halmos' words, we learned nothing at all, seems a case of throwing the baby out with the bath water.

---

[30] How often does one not read something like "The details of the proof are left to the reader", because it is considered that the techniques needed for the proof are known?

[31] The reviewing of computer-assisted proofs is indeed a difficult problem that has become very apparent in the context of Hales' proof of the sphere packing problem. In this respect it is interesting to read the account of the discussion between Robert MacPherson (editor of the *Annals of Mathematics*) and John H. Conway about adding a disclaimer or endorsement to the published proof in (Krantz, 2011, pp. 168–9). Eventually, neither an editorial disclaimer nor an editorial endorsement were added.

Clearly, because of the problem of unsurveyablity, it is indeed impossible to understand every detail of the proof. However, this does not exclude understanding on a more general level, on the level of the structure of the proof and the feasibility of the methods used. In order to understand this, we need to return to the actual proofs.

As was explained in §2.2 the Busy Beaver proofs discussed proceed in two main stages. The first stage concerns the reduction of the number of Turing machines to be considered by implementing certain theoretical considerations, normalization being the main technique. Clearly, these theoretical methods are explained and one can understand why they work. Although this first set of reductions is achieved by the computer, one cannot say that, given a specific machine that has been eliminated, one cannot understand why it has been eliminated. Furthermore, although one does not have every one of the details in order to understand why only 603,712 and not 603,711 holdouts remain in case of Kopp's tree normalization program, one can understand how the program works and why so many machines are eliminated.

In the second stage of the proof the idea is to differentiate between halting and non-halting machines by using several different heuristic programs. First, a program is used that runs each of the holdouts for some hundred of steps and eliminates those that have halted. Then the proof goes on with the remaining machines, trying to prove that each one of them is a non-halting machine through the detection of infinite loops. Now, in each of the proofs discussed, the remaining machines 'happened to be' cases of non-halting machines. Why? Because, through a complex interaction between man and the computer, it was found that all of these holdouts are cases of one of the different types of infinite loops. Clearly, this does not provide a good understanding. However, one can easily understand the general idea of the second stage of the proof. One can understand why all the remaining holdouts must be non-halting machines: during a process of exploration several different types of patterns were found. It can be proven that if a given Turing machine shows this patterns in its behavior then it will never halt. Now, for each of the holdouts it was proven—some by hand, most through the use of heuristic infinite-loop detection programs—that they are each cases of one of these types of patterns. Hence it follows that they cannot halt.

Of course, we do not have the details to understand why exactly 417 (and not say 320) of the 4-state machines, and only those, were identified as counters by Kopp's counter-detection program, nor do we have a complete understanding of why some machine x results in, say, a Christmas tree. However, we do know what a counter is, we do know why a counter must be a non-halting machine, we do know why the counter-detection program eliminates some machines and others not (even though these might still be counters), etc. In other words, although we do not have all the details and we do not have a kind of complete and detailed explanation or understanding of the Busy Beaver results, one cannot neglect that the papers on the Busy Beaver proofs do convey an understanding of how the proofs work and why they work on the global level. Furthermore, the proofs also give an insight on another level: they say something about the possible behavior one can expect from very simple programs, which are on the borderline of undecidability. This observation can be used in other settings. In fact, this is done in the paper by Machlin and Stout: the techniques and observations of the Busy Beaver proof are also used in the context of determining halting probabilities.

It is true that the Busy Beaver proofs do not give the kind of insight one gets from the several proofs of for instance the Pythagorean theorem. However, to call them oracles which do not provide any insight or understanding whatsoever is equally wrong.

4 Discussion

What exactly is the impact of the computer on mathematics? Has the computer really changed mathematical knowledge and the way we attain it, and, if yes, how? As was explained throughout the paper, this is an intricate question which has had no definite answer yet. A lot depends on one's own epistemological and ontological position. Questions like: "Are computer-assisted proofs fundamentally different from the usual proofs of mathematics?" or "Do computer-assisted proofs show that mathematical knowledge can be a posteriori?" are fundamental issues, issues that, in my opinion, cannot be answered satisfactorily on the basis of one micro-analysis. However, whatever one's answer might be to these fundamental issues, there is one thing which cannot be denied on the basis of this and other case studies. Computers are changing the way we are doing mathematics. They are changing 'mathematical practice'.

As is clear from the case-analysis provided here, computer-assisted proofs share some very typical features, problems and techniques. First of all, the aspect of man-computer interaction should not be underestimated. This is something completely new in mathematics: the fact that a proof is the result from a collaboration between a human and a non-human. It is the first time in history that a non-human is actively involved in the process of the proof. Although this non-human is mostly regarded as a mere quantitative help, it has led to fundamental qualitative changes, witness the fact that the mere increase in speed and memory due to the computer has made it possible to prove theorems that could not be or have not yet been proven without it. Besides this, the use of the computer almost naturally introduces the use of experimental methods in the process of finding a proof. These methods involve both the computer and the human, and arise from the complex interaction between man and machine.

A consequence of the involvement of the computer is that the representation of computer-assisted proofs, the way they are communicated, is different from that of usual proofs. A proof-as-communicated is no longer a sequence of lines of symbols that represent the proof found by the mathematician. It also includes a description of what the non-human has done (the computational process), what kind of results it communicates back (the output) and how the human has communicated his/her questions to the non-human (the code). Furthermore, unlike more traditional proofs, there is a strong emphasis on the process of finding a proof in the published papers of the computer-assisted proofs discussed in this paper, making explicit mention of the several experimental methods used, the complexity of this process, etc. As was explained in §2.3, one of the reasons for doing so is to deal on a local level with the problem of error and to allow for the reader-mathematician, even if he/she has not all the details, to follow the proof and verify it. Indeed, it is a typical property of computer-assisted proofs that they all suffer from the same fundamentally philosophical problems of unsurveyability, understanding and error. These problems seem unavoidable, but are dealt with on the local level. I.e., specific strategies are implemented both in the process of finding the proof as well as in the process of finding a good form for the the proof in order to 'manage' these problems.

The least one can thus conclude is that the process of finding a computer-assisted proof, the method of writing it down as well as the communication of a computer-assisted proof is different from those of usual proofs, at least, from a practical point of view. If one furthermore accepts that the way a proof is attained, its written-down form and the way it is communicated, is a determining feature of what a proof 'really' is, or better, means, then this alone changes (the meaning of) proof.

It is thus clear that, practically speaking, computer-assisted proofs do have an impact on mathematical practice. However, as long as computer-assisted proofs, and, more generally, computer-assisted mathematics, are the exception rather than the rule, the impact of the computer on mathematical practice remains rather limited. They only have a local impact, not a global one. The fact that proofs-made-flesh are different from the usual proofs of mathematics, the fact that one needs to deal with the fundamental problems of unsurveyability, understanding and error, the fact that experimental methods are needed and all the consequences these features and problems have—all these facts and effects can be discarded on the basis of the marginal role computer-assisted proofs play. The impact remains limited to a few cases and one can simply classify these cases as some of those exceptional 'behemoths' mathematics sometimes gives rise to. However, the question is whether this will indeed remain the case. It is only since the last 20 to 30 years that the computer and thus computer-time has become widely available to every mathematician, not only physically, in the sense that every mathematician now has his/her personal computer, but also intellectually, since more and more mathematicians are getting used to using computers. Furthermore, memory and computation speed are now exponentially larger than those of the early computers.[32] This wide availability of the computer and the increase in speed and memory, mere quantitative facts, could result and are in fact already resulting in more and more computer-aided mathematical results.

If the use of computers in mathematics is becoming more general, one cannot but conclude that they have an important impact on mathematics-as-practiced on the global level. A consequence of this is that, as the present case-analysis shows, the mathematicians will have to develop new methods and they will have to deal with the problems computer-assisted mathematics gives rise to. New results will be found as a consequence.

Is this where the impact of the computer stops? It is my view that as changes on the micro-level of the practice become more and more widespread, fundamental changes on the macro level and thus also changes in the foundations and philosophy of mathematics, become unavoidable. The way mathematics is perceived and understood changes. Questions concerning the certainty of mathematical knowledge, the fallible character of mathematics, etc. will only gain in importance, as more and more mathematicians are confronted with the consequences of computer-assisted mathematics. The fact that one is no longer able to check a proof completely, the fact that part of the proof is done by a non-human, the necessity of using experimental methods, the fact that one is confronted with very long proofs, etc. are real problems that not only change mathematics-as-practice but also the standard of what a proof is/should be and thus, ultimately, the standard of what mathematical knowledge is/should be.

Bibliography

Appel, K. and Haken, W. (1977). Every planar map is four colorable. Part I. Discharging. *Illinois Journal of Mathematics*, 21:429–490.

Appel, K. and Haken, W. (1986). The four color proof suffices. *Mathematical Intelligencer*, 8(1):10–20.

---

[32]ENIAC for example did not even store a megabyte, and it could handle 'only' 1000 computations in one second.

Appel, K., Haken, W., and Koch, J. (1977). Every planar map is four colorable. Part II. Reducibility. *Illinois Journal of Mathematics*, 21:491–567.

Arkoudas, K. and Bringsjord, S. (2007). Computer, justification and mathematical knowledge. *Minds and Machines*, 17(4):185–202.

Baker, A. (2008). Experimental mathematics. *Erkenntnis*, 68(3):331–344.

Bonsall, F. F. (1981). A down-to-earth view of mathematics. *American Mathematical Monthly*, 89(1):8–15.

Borwein, J. (2008). Implications of experimental mathematics for the philosophy of mathematics. In Gold, B. and Simons, R., editors, *Proof and other dilemmas: Mathematics and philosophy*, Spectrum, pages 33–60. Mathematical Association of America.

Borwein, J. and Bailey, D. (2003). *Mathematics by experiment. Plausible reasoning in the 21st century*. AK Peters, Wellesley, MA.

Borwein, J. and Bailey, D. (2004). *Experimentation in mathematics: Computational paths to discovery*. AK Peters, Wellesley, MA.

Brady, A. H. (1966). The conjectured highest scoring machines for Radó's $\sigma(k)$ for the value $k = 4$. *IEEE Transactions on Electronic Computers*, EC-15(5):802–803.

Brady, A. H. (1983). The determination of the value of Radó's noncomputable function σ for four-state Turing machines. *Mathematics of Computation*, 40(162):647–665.

Burge, T. (1998). Computer proof, apriori knowledge, and other minds. The sixth *Philosophical Perspectives* lecture. *Noûs*, 32(S12):1–37.

Calude, A. S. (2001). The journey of the four colour theorem through time. *New Zealand Mathematics Magazine*, 38(3):27–35.

De Mol, L. (2005). Study of fractals derived from IFS-fractals through metric procedures. *Fractals*, 13(3):237–244.

Detlefsen, M. and Luker, M. (1980). The four-color theorem and mathematical proof. *Journal of Philosophy*, 77(12):803–820.

Ekhad, S. B. (1990). A very short proof of Dixon's theorem. *Journal of Combinatorial Theory, Series A*, 54:141–142.

Gonthier, G. (2004). A computer-checked proof of the Four Color Theorem. Unpublished.

Hales, T. C. (2005). A proof of the Kepler conjecture. *Annals of Mathematics*, 162(3):1063–1183.

Halmos, P. R. (1990). Has progress in mathematics slowed down? *American Mathematical Monthly*, 97(7):561–588.

Kopp, R. J. (1981). The busy beaver problem. Master's thesis, State University of New York at Binghamton.

Krakowski, I. (1980). The four color problem reconsidered. *Philosophical Studies*, 38:91–96.

Krantz, S. G. (2011). *The proof is in the pudding. The changing nature of mathematical proof.* Springer, New York NY.

Lakatos, I. (1976). *Proofs and refutations: The logic of mathematical discovery.* Cambridge University Press, Cambridge. Edited by J. Worrall and E. Zahar.

Lam, C. W. H. (1991). The search for a finite projective plane of order 10. *American Mathematical Monthly*, 98(4):305–318.

Lam, C. W. H., Thiel, L., and Swiercz, S. (1989). The non-existence of finite projective planes of order 10. *Canadian Journal of Mathematics*, 41:1117–1123.

Lehmer, D. H. (1963). Some high-speed logic. In *Experimental arithmetic, high speed computing and mathematics*, volume 15 of *Proceedings of Symposia in Applied Mathematics*, pages 141–376.

Lehmer, D. H. (1966). Mechanized mathematics. *Bulletin of the American Mathematical Society*, 72(5):739–750.

Lehmer, D. H. (1969). Computer technology applied to the theory of numbers. In Leveque, W., editor, *Studies in Number Theory*, volume 6 of *Studies in Mathematics*, pages 117–151.

Lehmer, D. H., Lehmer, E., Mills, W., and Selfridge, J. L. (1962). Machine proof of a theorem on cubic residues. *Mathematics of Computation*, 16(80):407–415.

Levin, M. (1981). On Tymocsko's argument for mathematical empiricism. *Philosophical Studies*, 39:79–86.

Lin, S. and Radó, T. (1965). Computer studies of Turing machine problems. *Journal of the ACM*, 12(2):196–212.

Machlin, R. and Stout, Q. F. (1990). The complex behaviour of simple machines. *Physica D*, 42:85–98.

MacKenzie, D. (1999). Slaying the kraken: The sociohistory of mathematical proof. *Social Studies of Science*, 29(1):7–60.

Marxen, H. and Buntrock, J. (1990). Attacking the busy beaver 5. *Bulletin of the EATCS*, 40:247–251.

McCune, W. (1997). Solution of the Robbins problem. *Journal of Automated Reasoning*, 19(3):263–276.

Michel, P. (2004). Small Turing Machines and generalized busy beaver competition. *Theoretical Computer Science*, 326(1–3):45–56.

Pólya, G. (1954). *Mathematics and plausible reasoning, vol. 1: Induction and analogy in mathematics.* Princeton University Press, Princeton, NJ.

Radó, T. (1962). On non-computable functions. *Bell System Technical Journal*, 41(3):877–884.

Radó, T. (1963). On a simple source for non-computable functions. In Fox, J., editor, *Mathematical Theory of Automata*, volume XII of *Microwave Research Institute Symposia Series*, pages 75–81. Polytechnic Press.

Robertson, N., Sanders, D. P., Seymour, P., and Thomas, R. (1997). The four-color theorem. *Journal of Combinatorial Theory, Series B*, 70(1):2–44.

Seiden, S. (1998). A manifesto for the computational method. *Theoretical Computer Science*, 282(2):381–395.

Shanker, S. G. (1986). The Appel-Haken solution of the the four-color problem. In Shanker, S. G., editor, *Ludwig Wittgenstein: Critical Assessments*, pages 395–412, London. Croom Helm.

Swart, E. R. (1980). The philosophical implications of the four color problem. *American Mathematical Monthly*, 87:697–707.

Teller, P. (1980). Computer proof. *Journal of Philosophy*, 77(12):797–803.

Tymoczko, T. (1979). The four-color problem and its philosophical significance. *Journal of Philosophy*, 76(2):57–83.

Van Bendegem, J. P. (2005). The Collatz-conjecture. A case study in mathematical problem solving. *Logic and Logical Philosophy*, 14(1):7–23.

Van Kerkhove, B. and Van Bendegem, J. P. (2008). Pi on earth, or mathematics in the real world. *Erkenntnis*, 68(3):421–435.

Zeilberger, D. (1993). Theorem for a price: Tomorrows semi-rigorous mathematical culture. *Notices of the American Mathematical Society*, 40:978–981.

FotFS VII

François, K., Löwe, B., Müller, T., Van Kerkhove, B., editors,
Foundations of the Formal Sciences VII
Bringing together Philosophy and Sociology of Science

Sources for myths about mathematics. On the significance of the difference between finished mathematics and mathematics-in-the-making

CHRISTIAN GREIFFENHAGEN AND WES SHARROCK[*]

School of Social Sciences, Sociology, Arthur Lewis Building, University of Manchester, Manchester M13 9PL, United Kingdom
E-mail: {christian.greiffenhagen,wes.sharrock}@manchester.ac.uk

1 Introduction

Mathematical knowledge is often seen as a special kind of knowledge, namely as the very paradigm of certainty, universality, objectivity, or exactness. To some, it seems that in contrast to other forms of knowledge—including natural scientific knowledge—mathematical results are *absolutely* certain, indubitable, and infallible. For example, Pythagoras's Theorem was established over two millennia ago and still holds today. What other forms of activity can claim such definitive achievements?

> Whenever someone wants an example of certitude and exactness of reasoning, he appeals to mathematics. (Kline, 1980, p. 4)

> Mathematical results seem to be the paradigms of precision, rigor and certainty—from elementary theorems about numbers and geometric figures to the complex constructions of functional analysis and set theory. (Tymoczko, 1986, p. xiii)

> [...] mathematical knowledge is timeless, although we may discover new theories and truths to add; it is superhuman and ahistorical, for

---
[*]We are immensely grateful to the lecturers and students who allowed us to observe and record the lectures and supervision meetings and who more generally helped with the research. Christian Greiffenhagen gratefully acknowledges the support through a British Academy Postdoctoral Fellowship and a Simon Research Fellowship (funded through an endowment made to the University of Manchester).

Received by the editors: 29 May 2009; 11 December 2009; 12 January 2011.
Accepted for publication: 18 March 2011.

> the history of mathematics is irrelevant to the nature and justification of mathematical knowledge; it is pure isolated knowledge, which happens to be useful because of its universal validity; it is value-free and culture-free, for the same reason. (Ernest, 1996, p. 807)

> What is that distinguishes mathematical knowledge from other forms of knowledge? Students of all ages, whether they are studying algebra, trigonometry, geometry, or calculus, would agree on one universal theme: Mathematics is beyond doubt. A mathematical statement is either right or wrong, and any true statement can be proven. Moreover, once it is proved true, it can never later turn out to be false. (Henrion, 1997, p. 235)

Mathematical reasoning can also seem to constitute a special form of reasoning, which is encapsulated in mathematical proofs. Since proofs have a logical, strict, and formal character, it may be that the reasoning that creates them shares these characteristics. Mathematical reasoning has thus been likened to the functioning of a machine, which relentlessly follows a set of (mechanical) rules. Rota has called this the "machine grinder" conception, which conceives of mathematicians as driven by logical necessity (rather than imagination) and thus wholly mechanical in their "grinding out", solely according to the dictates of formal logic, successor proposition after successor proposition until arriving at a seemingly irresistible conclusion:

> According to this myth, the process of reasoning is viewed as the functioning of a vending machine which, by setting into motion a complex mechanism reminiscent of those we saw in Charlie Chaplin's film *Modern Times*, grinds out solutions to problems, like so many Hershey bars. (Rota, 1991, p. 175)

However, are these views of mathematics actually correct or are they only myths? Though some apparently accept these myths, there is also much scepticism about them, and a variety of scholars from the humanities and social sciences have challenged these myths, denying the absolute certainty of mathematics and the formal character of mathematical reasoning.

Historians of mathematics have pointed to disagreements and controversies in the history of mathematics, some going so far as to argue that mathematics—like the experimental sciences—goes through Kuhnian "revolutions" (Gillies, 1992). The most influential contribution has been Lakatos's (1976) *Proofs and Refutations*, in which Lakatos demonstrates that mathematical concepts (such as "polyhedron") are not permanently fixed, but can have a history of continuous modification. The same argument has been applied to the very standards and forms of mathematical proof, which have been shown not to be eternally fixed but subject to historical and cultural changes (Kleiner, 1991; MacKenzie, 2001).

A variety of philosophers have questioned the status of mathematical knowledge as *absolutely* certain (e.g., Quine, 1960; Wittgenstein, 1978; Lakatos, 1976; Kitcher, 1983; Tymoczko, 1986; Maddy, 1990). Often they have done so by drawing philosophical conclusions from specific mathematical developments, in particular, the discovery of non-Euclidean geometries at the end of the 19th century and the production of Gödel's Incompleteness Theorems in the early 20th century. Non-Euclidean geometries brought into question the view that the certainty of mathematics resides in its correspondence to reality, while Gödel's Incompleteness Theorems shook the belief that it would be possible to finally (and formally) prove the consistency of mathematics from a finite number of assumptions.

Rather than emphasizing the certainty of mathematics, researchers have thus started to talk of mathematics as "fallible", "uncertain", and "contingent" (e.g., Lakatos, 1976; Kline, 1980; Davis and Hersh, 1981; Ernest, 1998). Mathematical proofs are—and remain—open to revision (as is evidenced by the fact that some of them have been revised) and therefore cannot have been statements of absolute and eternal truths (since these could not possibly be questioned). This has led some researchers to argue that the belief in the (absolute) certainty of mathematics is only a reflection of (various) *myths* about mathematics. For example, Davis and Hersh (1981, p. 322) argue that the traditional picture of mathematics constitutes "the Euclid myth"; Hersh (1991, p. 130) maintains that myths about the unity, objectivity, universality, and certainty are widely spread; Dowling (2001, p. 21) also claims that many people subscribe to the "myth of certainty", which for Borba and Skovsmose (1997) actually constitutes an "ideology of certainty".

However, despite the fact that the "loss of certainty" (Kline, 1980) occurred over a century ago, it is also assumed by researchers that these myths *still* persist and that belief in them is widespread. Some researchers have tried to explain the existence of these myths by arguing that they originate in the way that mathematics has been, and still is, *presented and taught* in journals, textbooks, lectures, and classrooms. That is to say, it may be that mathematical knowledge is communicated and taught *so as to make it seem* that it is absolutely certain. The impression that mathematical results are certain may thus be a result of the fact that mathematical publications typically contain only the final results of mathematical investigations, i.e., typically do not mention all the things that didn't work out, the reasons why certain things were tried, or the ways in which the final theorem was modified during the course of proving. In particular, the modern *definition-theorem-proof* format of presenting mathematics (cf. Davis and Hersh, 1981, p. 151; Thurston, 1994, p. 163; Weber, 2004, p. 116) could be said to misrepresent the order in which mathematicians initially work out their proofs (if the format of definition-proof-theorem is to be understood as giving an order of presentation that reproduces the order of the mathematician's investigative reasoning, i.e., as suggesting that the mathematician started with clearly stated definitions, used those to form a conjecture, and then just wrote out the successive steps of the proof that are logically implied in the initial definition unhesitatingly one after the other).

Borasi (1992, p. 161), for example, suspects that people may mistake the way that mathematical results are presented as an account of how they were found:

> Perhaps because textbooks and lectures tend to present mathematical results in a "neat" and organized way, few people realize that those results have not always been achieved in a straightforward manner. (Borasi, 1992, p. 161)

Crawford et al. (1998, p. 466) argue that the way that mathematics is presented to students hides important information and thereby misrepresents the nature of mathematics:

> Most students learn mathematics at school and university in a competitive environment where mathematics is presented as a finished and polished product and where the assessment encourages students to reproduce authoritative statements of fact [...]. In presenting mathematics in this way, students are provided with mathematical information about concepts, proofs, techniques and skills, but the processes which created this information are hidden [...]. The lack of awareness of these creative processes makes it difficult for students to experience mathematics as personally meaningful and misrepresents the nature of mathematics itself.

Powell and Brantlinger (2008, p. 428) argue that "the struggle of discovery is [...] usually missing from the narratives [students] read in mathematics texts". Livingston (2006, p. 60) wonders whether since proofs only exhibit "the reasoning that someone, or some others, had discovered, but not how they had gone about discovering it", students may mistake "the logic of a proof [as] the justification of how it was found". Similar considerations lead Campbell to speak of the "curse of Euclid":

> Euclid's axiomatic procedure is a breakthrough; it is a procedure for the unification of material. It allows key assumptions to stand out. It allows for systematic procedures of verification. But so long as students are misled into believing that the polished jewels are the actual reasoning rather than the end product of reasoning, just so long will it be that Euclidean geometry will remain a curse rather than a blessing to the teaching of reasoning. (Campbell, 1976, p. 342)

Ernest suggests that textbooks therefore constitute a "pedagogical falsification" of mathematical reasoning:

> Lakatos (1976) and others have criticized the pedagogical falsification perpetrated by the standard practice of presenting advanced learners with the sanitized outcomes of mathematical enquiry. Typically advanced mathematics text books conceal the processes of knowledge construction by inverting or radically modifying the sequence of transformations used in mathematical invention, for presentational purposes. The outcome may be elegant texts meant for public consumption, but they also generate learning obstacles through this reformulation and inversion. (Ernest, 2008, p. 67)

Overall, these quotations demonstrate a common—though vague and variable—sense that the way that mathematics is presented and written, in particular, the stress on established results (emphasising certainty) and on proofs (emphasising formal rules), may be an important source for the myth of mathematics as absolutely certain and as the province of special kinds of individuals with distinctive powers of reason. These impressions may be reinforced by the style in which mathematical reports are written, which could be described as "impersonal", "objective", or "authoritative" (Davis and Hersh, 1981, p. 36; Morgan, 1998, p. 11; Burton and Morgan, 2000, p. 435) and which could be seen to obliterate the understanding that proofs must be somebody's creation (thus conveying the impression that mathematics is a superhuman achievement with results appearing as if "untouched by human hand").

In order to demythologize mathematics, researchers have contrasted the way that mathematics is written and presented with how it is actually done and practiced. In other words, they have pointed to the differences between finished mathematics (which can be found in textbooks or lectures) and mathematics-in-the-making (when researchers are still in the process of coming up with the various definitions, theorems, and proofs). It has been argued that when the attention is turned to mathematics-in-the-making, a very different view of mathematics emerges, one that is much less alienating and therefore more welcoming to outsiders. This argumentative strategy is nicely captured in an article by Reuben Hersh.

2 Mathematics as divided into a front and a back

In *Mathematics has a front and a back* (1991), Hersh adopts Goffman's (1959) dramaturgical model of social establishments as divided between a "front" and a "back" region to mathematics.[1] In Goffman's terms, different areas of social establishments are separated by the need to project and sustain public impressions. A (usually more elevated) image is projected by conduct in the "front" region, since it is this area to which the audience is allowed access. However, what goes on in the "front" is supported by

---
[1] Cf. also Hersh (1997, pp. 35–39).

"backstage" activities, which include all sorts of things that are incongruous with (and usually diminishing of) the image being projected by the front and from which the audience—its public—is excluded. In a restaurant, for example, the "front" is the dining area where customers are offered hygienically presented dishes, which, of course, are the product of various (perhaps unhygienic) activities in the kitchen "backstage". Furthermore, while the dining area is often quiet, neat, and tidy, the kitchen is smelly, noisy, and chaotic (and there may be real discrepancies between the politely subordinate conduct of waiting staff in view of the customers and the disrespect, even contempt, on show backstage).

Hersh uses Goffman's conception of the "front" and the "back" to distinguish between the products of mathematical activities (finished mathematics) and the processes of mathematics (mathematics-in-the-making):

> [...] the "front" of mathematics is mathematics in "finished" form, as it is presented to the public in classroom, textbooks, and journals. The "back" would be mathematics as it appears among working mathematicians, in informal settings, told to one another in an office behind closed doors.
>
> Compared to "backstage" mathematics, "front" mathematics is formal, precise, ordered and abstract. It is separated clearly into definitions, theorems, and remarks. To every question there is an answer, or at least, a conspicuous label: "open question". The goal is stated at the beginning of each chapter, and attained at the end.
>
> Compared to "front" mathematics, mathematics "in back" is fragmentary, informal, intuitive, tentative. We try this or that, we say "maybe" or "it looks like". (Hersh, 1991, p. 128)

Not only are there differences between the "front" and the "back", but the audience is typically not allowed access to the "back". In other words, the activities that go on in the back are being kept from the audience:

> So its "front" and "back" will be particular kinds or aspects of mathematical activity, the public and private, or the part offered to "outsiders" (down front) versus the part normally restricted to "insiders" (backstage). (Hersh, 1991, p. 128)

Hersh takes the separation into a "front" and "back" to be the source for the various myths associated with mathematics. Although mathematicians themselves may well know that their work does not conform to this "official" image of mathematics,[2] they nonetheless sustain the mythology through their publication arrangements and educational practices:

---

[2]Other authors suppose that mathematicians themselves may have swallowed the myths or ideology about their own discipline.

> [...] the front/back separation makes possible the preservation of a myth [...]. By a myth we shall mean simply taking the performance seen from up front at face value; failing to be aware that the performance seen "up front" is created or concocted "behind the scenes" in back. This myth, in many cases, adds to the customer's enjoyment of the performance; it may even be essential. [...] Mathematics, too, has it myths. One of the unwritten criteria separating the professional from the amateur, the insider from the outsider, is that the outsiders are taken in (deceived), the insiders are not taken in. (Hersh, 1991, p. 129)

And the point of these myths is to support the social institution of mathematics:

> By calling these beliefs myths, I am not declaring them to be false. A myth need not be false to be a myth. The point is that it serves to support or validate some social institution [...].
>
> [...] the unity, universality, objectivity, and certainty of mathematics are beliefs that support and justify the institution of mathematics. (For mathematics, which is an art and a science, is also an institution, with budgets, administrations, publications, conferences, rank, status, awards, grants, etc.)
>
> Part of the job of preparing mathematics for public presentation— in print or in person—is to get rid of all the loose ends. If there is disagreement whether a theorem has really been proved, then that theorem will not be included in the text or the lecture course. The standard style of expounding mathematics purges it of the personal, the controversial, and the tentative, producing a work that acknowledges little trace of humanity, either in the creators or the consumers. This style is the mathematical version of "the front".
>
> Without it, the myths would lose much of their aura. If mathematics were presented in the same style in which it is created, few would believe in its universality, unity, certainty, or objectivity. (Hersh, 1991, p. 130–131)

There are a number of important aspects to Hersh's adaption of Goffman's front/back dramaturgical model to mathematics:

Firstly, Hersh suggests a *strong* segregation between how mathematics is presented "in the front" and how it is actually done "in the back", insinuating that there might be two different kinds of mathematics (or mathematical reasoning), one "formal" and "logical", the other "informal" and "chaotic". As Ernest (2008, p. 66) restates Hersh:

> As Hersh [...] has pointed out, mathematics (like the restaurant or theatre) has a front and a back. What is displayed in the front for

public viewing is tidied up according to strict norms of acceptability, whereas the back (where the preparatory work is done) is often messy and chaotic.

Secondly, Hersh clearly implies that this segregation of things done "in the back" that are not stated in "front" presentations is a *deliberate* and *calculated* move to keep that information from the audience:

> The purpose of a separation between front and back is not just to keep the customers from interfering with the cooking, it is also to keep the customers from knowing too much about the cooking. (Hersh, 1991, p. 129)

Thirdly, Hersh assumes that these activities are concealed or hidden from the audience in the interest of *protecting the institutional interests of mathematics* (because public awareness of them would transform, even devalue, the high status assessment that allegedly follows from the "front" presentations). Hersh speaks of the backstage activities as going on "behind closed doors", which suggests that what goes on there would be embarrassing and would detract from the merits of mathematics if made public. Henrion (1997, p. 249) similarly states:

> One reason that the mathematics community is not more active in conveying a more accurate picture of mathematics is that much of the power and prestige of mathematics comes from its claim to certainty and its image as an "exact science".

The main thrust of Hersh's front/back imagery is that mathematical results come to be treated by "the audience" or "the public" *as if* they could not possibly be questioned, which gives mathematicians an awesome authority. Those outside the mathematical community seem to have no choice but to be obedient to whatever (mathematical) directives mathematicians give them.[3] Of course, no one supposes that this authority derives from the personal characteristics of (balding, spotty, aging, shabby) mathematicians, so it must stem from the nature of the mathematics itself. Hersh suggests that the idea that mathematicians can't possibly make mistakes is—at least partially—a result of the way in which they present their work.

---

[3] According to David Bloor (e.g., 1976, 1994) many people (falsely) believe in the universality of mathematics, i.e., assume that 2 + 2 must equal 4, and therefore do not realize the conventional character of mathematics (for example, the use of a decimal rather than binary notation). The aim of Bloor's sociology of mathematics is thus to demonstrate that mathematical propositions "could have been otherwise" and thereby to show that no one is compelled to accept these propositions on the grounds that they could not possibly be otherwise. These views are questioned in Greiffenhagen and Sharrock (2006, 2009).

The front/back scheme is invoked to debunk myths about mathematics by showing that mathematical results and reasoning are less certain and formal than they appear to be. This is done by showing that the reality of mathematical practice relevantly deviates from its frontal presentation. Two main strategies are at play. Firstly, it is held that the "front" evades mention of mathematician's controversies or doubts about their own or others' work (since preparing things for publication includes "to get rid of all the loose ends" and to "purge [...] it of the personal and controversial, and the tentative"). Secondly, it is claimed that the course of a mathematical reasoning is not an inexorable progression of inevitable steps. In other words, mathematicians reason heuristically (cf. Pólya, 1945, 1954a,b) and often engage in revision of their strategies for forming their sought after proof. Again, Hersh suggests that because publications do not declare the tentativeness, uncertainties, and informality associated with working on proofs, they are occluding the fact that mathematical work is "contingent", "revisable", and "fallible". The upshot of Hersh's argument seem s to be that if the audience was admitted or given access to the "back" of mathematics—as, for example, attempted in *The Mathematical Experience* (Davis and Hersh, 1981) or *What is Mathematics, Really?* (Hersh, 1997)—they could be emancipated from the mythologized picture of mathematics.

We are certainly not going to dispute that myths about mathematics do circulate or that there are manifest differences between finished mathematics and mathematics-in-the-making. Neither do we address the "internal" question of whether the historical shifts in ways that mathematics has been written and presented have been an advantage or disadvantage to the discipline. Mathematicians of course debate the "best" way of presenting mathematical results (cf. Ulam, 1976, pp. 276–277): Is it good to have "succinct" papers that highlight the important points, or is it better to have "complete" papers that deal with every detail? Should one start with a short introduction that explains the motivations of trying to tackle the problem? Should one just state the theorem and then present the proof? Burton and Morgan (2000, p. 449), for example, quote a mathematician who calls for changes:

> I get annoyed with some of my collaborators and a lot of the papers I am sent, which are definition, theorem, lemma, proof. That seems to me to be appallingly bad. It is the sort of thing that no one is ever going to want to read. I think it is important to grab the reader from the opening sentence. Not "Let A be a class of algebras such that ..." Change it to "This paper opens a new chapter in duality theory."

Whatever the relevant pros and cons, these "internal" debates are different from the "external" question of whether the way in which mathematical papers are written implicates mythical conceptions of mathematics (although there are connections insofar as styles are condemned for being off-putting for imagined readerships). What we want to question is Hersh's suggestion that looking at mathematics "in the front" or "in the back" yields two incongruent views of mathematics. We would therefore like to take a closer look at two examples of mathematical practice, one taken from the "front" and one from the "back".

3 Towards a sociology of mathematics

It is surprising how very few detailed accounts of what mathematicians do *either* in the "front" *or* in the "back" can be found in the literature. The supposed features of "the front" as well as "the back" are typically not derived from the close analysis of specific examples (e.g., a particular lecture, textbook, or journal paper) or a corpus of them. This in turn makes it difficult to check whether the descriptions of the "front" or "back" are adequate.

There have been a number of very illuminating studies in the history of mathematics (e.g., Lakatos, 1976; MacKenzie, 1999, 2001; Netz, 1999, 2009; Warwick, 2003), which have in particular demonstrated the changing nature of many mathematical concepts (including the very notion of what constitutes a mathematical proof). However, there are very few studies that are based on observations of people engaged in doing mathematics. Even the recent sociological studies of mathematical disciplines (e.g., Livingston, 1986, 1999, 2006; Merz and Knorr-Cetina, 1997; Rosental, 2003, 2008—cf. also Greiffenhagen, 2008, 2010) are in no way straightforward ethnographies comparable to the laboratory studies of the experimental sciences. Part of the reason may be that it is simply more difficult to do such studies in the case of "theoretical" or "conceptual" practice:

> It is easy to study laboratory practices because they are so heavily equipped, so evidently collective, so obviously material, so clearly situated in specific times and spaces, so hesitant and costly. But the same is not true of mathematical practices: notions such as "demonstration", "modelling", "proving", "calculating", "formalism", "abstraction" resist being shifted from the role of indisputable resources to that of inspectable and accountable topics. It is as if we had no tool for holding such notions under our eyes for more than a fleeting moment, or simply no metalanguage with which to register them. (Latour, 2008, p. 444)

For the past six years, we have been conducting sociological studies of mathematics, trying to find "perspicuous settings" in which different features of mathematical practice become observable. We decided to focus on situations in which mathematicians come together to discuss mathematics. In a first study we observed and recorded three different graduate courses in mathematical logic for several months. In these lectures, an experienced mathematician demonstrates mathematical expertise to novices, showing what is treated as important and noteworthy about the "archive" of mathematics. In a second study we attended for over a year the (almost) weekly meetings between a supervisor and his doctoral students where they discuss the problems that the student has been working on.

In our first study, we video-taped different lecturers giving graduate lectures in mathematical logic (Figure 1). We do not here have the space to give a detailed example (which we plan to do in a future paper), but can only make some preliminary observations with respect to what one can see when one looks closely what happens in mathematical lectures.

The lectures that we observed followed the typical *definition-theorem-proof* format of presenting mathematics. Furthermore, like many lectures in mathematics, what the lecturer wrote on the board to a large part corresponded to the script that had been handed to students at the beginning of the course. These lectures thus are a perspicuous example of mathematics in the "front", since what is communicated

> is formal, precise, ordered and abstract. It is separated clearly into definitions, theorems, and remarks. To every question there is an answer, or at least, a conspicuous label: "open question". The goal is stated at the beginning of each chapter, and attained at the end. (Hersh, 1991, p. 128)

However, in which sense does this style of presenting mathematics contribute to myths about mathematics?

The first thing to notice is that lecturers did not simply copy the text from the script to the board. Rather, lecturers spent the majority of the time "working through" various proofs, typically without looking into the script (except on a few occasions when they wanted to check a particular detail or "got lost"), making a lot of additional comments, for example, highlighting which steps in the proof were important or noteworthy, how the steps depend on results established earlier, where the assumptions in the theorem were used in the proof, and so on.

This at least partly explains why lecturers recited theorems and proofs to students when the students could read them for themselves in the handouts of the course. Learning mathematics does involve reading a lot of theorems and proofs, but such reading requires work. Both lecturers and students are aware that reading through a complex proof (once) does not equip students with an (adequate) understanding of it. In that sense, a lecture is only the first step in a process. Lecturers expect that students will spend additional hours of individual study, together with attempting prescribed exercises, in order to understand the materials covered in the lecture (lecturers also expect that students are likely to have initial difficulties of understanding).

FIGURE 1. Graduate lectures

As Davis and Hersh (1981, p. 281) observe, a mathematical proof is only superficially comparable to a notated musical score and therefore only seems to be accessible to sight reading. However, this is something that students are *very* aware of. Although these students were still less than fully qualified practitioners, it is difficult to see how they would get a wrong picture of mathematics from the particular style in which the materials were presented in these lectures (a style, which does not differ from a first-year undergraduate course).

Furthermore, it is important to note that lecturers did not purport to give a report (in the form of a historical, sociological or even anecdotal description) on how the various theorems, definitions, or proofs were "found". Instead, lecturers "simply" exhibited how particular theorems or proofs "work" regardless of how they were found. The emphasis in these lectures was on intelligibility, not truthful historical reporting. The comments of lecturers were designed to make the various theorems and proofs intelligible to students, independently of knowing much, if any, biographical detail of the authors or anything about the extended and detailed work that went into developing the results. Although lecturers did not explicitly mention that the definitions, theorems, and proofs do not "fall from the sky" (but are the result of revisions, false avenues, etc.), this does not mean that they were "hiding" this from students. Students may have been frustrated that they failed to understand how a particular proof works or that they failed to solve some of the exercise problems. Even so, it would be strange to suppose that students as a consequence of the "formal, precise, ordered and abstract" way in which the results were presented in these lectures believed that it was possible for the originators of these results to achieve them without any effort or by applying a "mechanistic" procedure.

In sum, a careful consideration of looking at a concrete example of mathematics "in the front" (here: a mathematical lecture) shows that it does not necessarily lead to any of the myths about mathematics described by Hersh.

In a second study, we attended the weekly meetings between a supervisor and his doctoral students (Figure 2). In these sessions, the supervisor would discuss the work of the student, sometimes on the basis of some materials that the student had sent to the supervisor prior to the meeting. At other times, the student would provide an oral account of what he had been working on—which typically involved explaining why and where he did get "stuck" (cf. Merz and Knorr-Cetina, 1997)—and the supervisor would respond by making various suggestions on how the student could proceed.

The discourse in these meetings was indeed very different from that in the lectures and constitutes a good example of mathematics "in the back", which is described by Hersh (1991, p. 128) as "fragmentary, informal, intuitive, tentative. We try this or that, we say 'maybe' or 'it looks like'".

In these sessions, neither the supervisor nor the student would write down fully worked out proofs on the board (in fact, if the student had submitted a proof which was deemed to be complete-for-all-practical-purposes they typically did *not* talk further about it, but the conversation would move on to what could be done next, i.e., how to make further progress building on what they had proved so far). Rather than presenting finished mathematics, the supervisor and doctoral student used the board as a focus for their discussions, a place to sketch out ideas, conjectures, hunches, reasons for trying this or that.

The main aim of their meetings was to try out ideas, fully aware that they would only be able to work them out initially and partially in the

FIGURE 2. Supervision meetings

meeting. Sometimes it transpired relatively quickly that an idea would definitely not work and should be abandoned. However, more often than not, the idea remained to be worked through more systematically and there remained uncertainty as to whether a promising idea would work out in detail. Another aim of these meetings thus was to make assessments with respect to which problems it would be worth pursuing. These researchers, like other practical decision makers, were sensitive to economising their investment of effort as well as with the pacing of their inquiries, and so the decision as to whether to attempt to construct a proof or to search for a counterexample was to be considered in terms of the estimated likelihood that one, rather than the other, would pay off, and the amount of time that would need to be invested in getting out the correct conclusion.

Their discussions in these sessions were often not decisive, but this did not mean that there was a permanent tentativeness about determining whether an idea would or would not work in detail. Further additional work would often give a firm verdict (although some ideas remained unresolved either way). Although their work could be described as "intuitive", this does not mean that what they were doing resembled anything like arbitrary guesswork. One of the dangers of attacking a deductivist picture of mathematics is to suggest the opposite extreme. Experienced mathematicians do make guesses and rely on intuition, but do this on the basis of a vast armoury of accepted techniques and tricks.

Although the conversation in this setting did not resemble the discourse in the lecture, it was constantly oriented towards it. The aim was to come up with theorems and proofs that *could* and, if successful, would be presented in the style of the lecture. Conversely, both researchers constantly made use of the results and techniques that were taught in the lectures, since these are amongst the stock resources of the field. In that sense, it is difficult to see how the "back" was strongly separated from the "front", since the composition of their new proof is done through—in part—the use of results and techniques drawn from already established proofs (i.e., their innovation "in the back" involved the application of what they had learned from studying proofs "in the front"). In other words, there was not any simple discontinuity between "rough ideas that are good enough for our mathematical purposes" and the "formal, tidy presentation necessary to meet the needs of the audience", because, as far as these researchers were concerned, if they only had a rough idea, they did not necessarily have anything that *could* be presented in the format of a lecture.

In sum, looking at a concrete example of mathematics "in the back" (here: a meeting between a supervisor and his doctoral student) shows that it is not fundamentally different from mathematics "in the front".

4 Conclusion

Both Goffman and Hersh seem to suggest that the separation of a setting into a "front" and "back" can serve as the basis of various myths. Goffman writes that in the "front"

> [...] errors and mistakes are often corrected before the performance takes place, while telltale signs that errors have been made and corrected are themselves concealed. In this way an impression of infallibility, so important in many presentations, is maintained. (Goffman, 1959, p. 52)

While Hersh (1991, p. 127) states that "[a]cceptance of these myths [unity, objectivity, universality, and certainty] is related to whether one is located in the front or the back" and that "[m]ainstream philosophy doesn't know that mathematics has a back" (Hersh, 1997, p. 36).

We would be foolish to pretend that questionable conceptions about mathematics do not circulate (especially within philosophy) or that they cannot change over time ("fallibilism" was not a term widely used even in the nineteenth century). At the same time, since amongst the mythologisers are included Russell, Hilbert, and Bourbaki (and formalists generally), all of whom knew how to do advanced mathematics, we do not see that it is a lack of exposure to backregions that makes people susceptible to these mythical conceptions. People may have wrong ideas about mathematics, but we have questioned whether these are a creation of external features of the way in which textbooks, lectures, and other publications are set out in public formats, and have suggested that there is no compelling reason to think they are.

We have been arguing that whatever the origin of the various myths of mathematics might be, it cannot be the necessity to segregate one form of mathematical reasoning (that which mathematicians present themselves as using) from another (that which mathematicians *really* employ to do mathematics) so as to prevent "the audience" becoming aware of the difference. Rather than being helpful, the application of the front/back metaphor exaggerates the discontinuities between mathematics-in-publications and mathematics-in-the-making. Whilst Hersh might laudably aim to demystify mathematics, there is in our view the risk that in the end he will only service re-mystification of it.

Bibliography

Bloor, D. (1976). *Knowledge and social imagery*. Routledge & Kegan Paul, London.

Bloor, D. (1994). What can the sociologist of knowledge say about 2 + 2 = 4? In Ernest, P., editor, *Mathematics, education and philosophy: An international perspective*, pages 21–32, London. Falmer.

Borasi, R. (1992). *Learning mathematics through inquiry*. Heinemann, Portsmouth, NH.

Borba, M. and Skovsmose, O. (1997). The ideology of certainty in mathematics education. *For the Learning of Mathematics*, 17(3):17–23.

Burton, L. and Morgan, C. (2000). Mathematicians writing. *Journal for Research in Mathematics Education*, 31(4):429–453.

Campbell, D. (1976). *The whole craft of number*. Prindle, Weber & Schmidt, Boston.

Crawford, K., Gordon, S., Nicholas, J., and Prosser, M. (1998). Qualitatively different experiences of learning mathematics at university. *Learning and Instruction*, 8(5):455–468.

Davis, P. J. and Hersh, R. (1981). *The mathematical experience*. Birkhäuser, Boston.

Dowling, P. (2001). Mathematics education in late modernity: Beyond myths and fragmentation. In Atweh, B., Forgasz, H., and Nebres, B., editors, *Sociocultural research on mathematics education: An international perspective*, pages 18–36, Mahwah, NJ. Lawrence Erlbaum.

Ernest, P. (1996). Popularization: myths, massmedia and modernism. In Bishop, A. J., Clements, K., Keitel, C., Kilpatrick, J., and Laborde, C., editors, *International handbook of mathematics education*, pages 785–817, Dordrecht. Kluwer.

Ernest, P. (1998). *Social constructivism as a philosophy of mathematics*. State University of New York Press, Albany, NY.

Ernest, P. (2008). Opening the mathematics text: What does it say? In de Freitas, E. and Nolan, K., editors, *Opening the research text: Critical insights and in(ter)ventions into mathematics education*, pages 65–80, Dordrecht. Springer.

Gillies, D., editor (1992). *Revolutions in mathematics*. Oxford University Press, Oxford.

Goffman, E. (1959). *The presentation of self in everyday life*. Doubleday, Garden City, New York.

Greiffenhagen, C. (2008). Video analysis of mathematical practice? Different attempts to 'open up' mathematics for sociological investigation. *Forum: Qualitative Social Research*, 9(3).

Greiffenhagen, C. (2010). A sociology of formal logic? Essay review of Claude Rosental's 'Weaving self-evidence'. *Social Studies of Science*, 40(3):471–480.

Greiffenhagen, C. and Sharrock, W. (2006). Mathematical relativism: logic, grammar, and arithmetic in cultural comparison. *Journal for the Theory of Social Behaviour*, 36(2):97–117.

Greiffenhagen, C. and Sharrock, W. (2009). Two concepts of attachment to rules. *Journal of Classical Sociology*, 9(4):405–427.

Henrion, C. (1997). *Women in mathematics: The addition of difference*. Race, Gender, and Science. Indiana University Press, Bloomington, IN.

Hersh, R. (1991). Mathematics has a front and a back. *Synthese*, 88(2):127–133.

Hersh, R. (1997). *What is mathematics, really?* Oxford University Press, Oxford.

Kitcher, P. (1983). *The nature of mathematical knowledge*. Oxford University Press, Oxford.

Kleiner, I. (1991). Rigor and proof in mathematics: A historical perspective. *Mathematics Magazine*, 64(5):291–314.

Kline, M. (1980). *Mathematics: The loss of certainty*. Oxford University Press, Oxford.

Lakatos, I. (1976). *Proofs and refutations: The logic of mathematical discovery*. Cambridge University Press, Cambridge. Edited by J. Worrall and E. Zahar.

Latour, B. (2008). The Netz-works of Greek deductions. *Social Studies of Science*, 38(3):441–459.

Livingston, E. (1986). *The ethnomethodological foundations of mathematics*. Routledge, London.

Livingston, E. (1999). Cultures of proving. *Social Studies of Science*, 29(6):867–888.

Livingston, E. (2006). The context of proving. *Social Studies of Science*, 36(1):39–68.

MacKenzie, D. (1999). Slaying the kraken: The sociohistory of mathematical proof. *Social Studies of Science*, 29(1):7–60.

MacKenzie, D. A. (2001). *Mechanizing proof: Computing, risk, and trust*. MIT Press, Cambridge, MA.

Maddy, P. (1990). *Realism in mathematics*. Clarendon Press, Oxford.

Merz, M. and Knorr-Cetina, K. (1997). Deconstruction in a 'thinking' science: Theoretical physicists at work. *Social Studies of Science*, 27(1):73–111.

Morgan, C. (1998). *Writing mathematically: The discourse of investigation*. Falmer, London.

Netz, R. (1999). *The shaping of deduction in Greek mathematics: A study in cognitive history*. Cambridge University Press, Cambridge.

Netz, R. (2009). *Ludic proof: Greek mathematics and the Alexandrian aesthetic*. Cambridge University Press, Cambridge.

Pólya, G. (1945). *How to solve it: A new aspect of mathematical method*. Princeton university Press, Princeton, NJ.

Pólya, G. (1954a). *Mathematics and plausible reasoning, vol. 1: Induction and analogy in mathematics*. Princeton University Press, Princeton, NJ.

Pólya, G. (1954b). *Mathematics and plausible reasoning, vol. 2: Patterns of plausible inference*. Princeton University Press, Princeton, NJ.

Powell, A. and Brantlinger, A. (2008). A pluralistic view of critical mathematics. In Matos, J. a. F., Valero, P., and Yasukawa, K., editors, *Proceedings of the Fifth International Mathematics Education and Society Conference Albufeira, Portugal, 16–21 February 2008*, pages 424–433, Lisbon. Universidade de Lisboa.

Quine, W. V. O. (1960). *Word & object*. MIT Press, Cambridge, MA.

Rosental, C. (2003). Certifying knowledge: The sociology of a logical theorem in artificial intelligence. *American Sociological Review*, 68(4):623–644.

Rosental, C. (2008). *Weaving self-evidence: A sociology of logic*. Princeton University Press, Princeton, NJ. Translated by C. Porter.

Rota, G.-C. (1991). The pernicious influence of mathematics upon philosophy. *Synthese*, 88(2):165–178.

Thurston, W. P. (1994). On proof and progress in mathematics. *Bulletin of the American Mathematical Society*, 30(2):161–177.

Tymoczko, T., editor (1986). *New directions in the philosophy of mathematics: An anthology*. Birkhäuser, Boston.

Ulam, S. M. (1976). *Adventures of a mathematician*. Scribner, New York, NY.

Warwick, A. (2003). *Masters of theory: Cambridge and the rise of mathematical physics*. University of Chicago Press, Chicago, IL.

Weber, K. (2004). Traditional instruction in advanced mathematics courses: A case study of one professor's lectures and proofs in an introductory real analysis course. *Journal of Mathematical Behavior*, 23(2):115–133.

Wittgenstein, L. (1978). *Remarks on the foundations of mathematics*. Basil Blackwell, Oxford, revised edition.

François, K., Löwe, B., Müller, T., Van Kerkhove, B., editors,
Foundations of the Formal Sciences VII
Bringing together Philosophy and Sociology of Science

On the curious historical coincidence of algebra and double-entry bookkeeping

ALBRECHT HEEFFER[*]

Centrum voor Logica en Wetenschapsfilosofie, Universiteit Gent, Blandijnberg 2, 9000 Gent, Belgium
E-mail: filidoor@gmail.com

1 Introduction

The emergence of symbolic algebra is probably the most important methodological innovation in mathematics since the Euclidean axiomatic method in geometry. Symbolic algebra accomplished much more than the introduction of symbols in mathematics. It allowed for the abstraction and generalization of the concepts of number, quantity and magnitude. It led to the acceptance of negative numbers and imaginary numbers. It gave rise to new mathematical objects and concepts such as a symbolic equation and an aggregate of linear equations, and revealed the relation between coefficients and roots. It allowed for an algebraic approach to ancient geometrical construction problems and gave birth to analytical geometry. Why did this important methodological revolution happen? Why did it happen in Europe and not in Asia while Indian and Chinese algebra were more advanced before the fourteenth century? Why did it happen in the European Renaissance?

We can only touch the surface of possible answers to these fundamental questions within the scope of this paper. However, we would like to argue that the answers will involve multiple disciplines and will go beyond the boundaries of the history of mathematics. Most historians have taken for granted that symbolic algebra was an inevitable step within the logical development of mathematics. But can we speak of a logic of historical necessity? The history of mathematics at least teaches us that there have been developments within mathematics that were not in logical sequence.

---

[*]This paper was written while the author was visiting researcher at the Center for Research in Mathematics Education at Khon Kaen University, Thailand.

Received by the editors: 10 February 2009; 7 July 2009.
Accepted for publication: 28 October 2009.

A full notion of the function concept was developed only after the calculus, while textbooks on calculus first introduce functions and then move to differentiation and integration. Some concepts emerged within a historical context where no sensible interpretation could be given to their meaning. A notorious example is imaginary numbers. Still, in historical accounts these anomalies and anachronisms are considered exceptions, and *exceptio probat regulam*. What if there is no such logic of historical development? Then all historical questions must be addressed within their cultural and social-economical context. Answers cannot be found by appealing to the next logical step in the development, but only in the relation between its practices and their meaning within the society. In other words, philosophy, history and sociology of science all contribute to possible answers to the questions we have raised.

In this text, we shall first give a short overview of the internalistic approach to the emergence of symbolic algebra which, as we shall show, is present in most historical accounts on the history of algebra. We shall then present some studies which take a contextual approach to developments in mathematics during the period we are addressing. We then present our own position that symbolic algebra was made possible by the central idea of value as an objective quantity in mercantilism. As an illustration of how important developments in mathematics can be matched with macroeconomical changes in society we draw the parallel between symbolic algebra and double-entry bookkeeping. These two developments of the fourteenth and fifteenth century were both instrumental in the objectivation of value and they supported the reciprocal relations of exchange on which mercantilism depended. To demonstrate our proposition, we shall present a case study to show how symbolic algebra and double-entry bookkeeping function in our understanding of a special class of bartering problems. It would be wrong to understand a socio-economical account of the history of mathematics as the right one or the only one. At the contrary, we believe that a pluralism of explanations leads to a better understanding. However, concerning the history of European algebra too much emphasis has been put on internal mechanisms and we present our account as complementary to these approaches.

2 Internalist accounts of the history of algebra

Algebra was introduced in medieval Europe through the Latin translations of Arabic texts between 1145 and 1250 and Fibonacci's *Liber Abbaci* (1202) (Boncompagni, 1857; Sigler, 2002). Algebraic problem solving was further practiced within the so-called abbaco tradition in cities of fourteenth- and fifteenth century Italy and the south of France.[1] From the sixteenth century, under the influence of the humanist program to provide new foundations to this *ars magna*, abbaco algebra evolved to a new logistics of species with François Viète (1591) as the key figure. With Descartes's *Geometry*, this new kind of algebra progressed into our current symbolic algebra. This is a brief characterization of the current view of scholars on the history of European symbolic algebra.

Most of the studies on the history of algebra provide an internalistic account. They accept implicitly or more explicitly that the development towards symbolic algebra was inevitable and depended on some internal mechanisms and intrinsic processes. Moritz Cantor whose *Vorlesungen* (1880; 1892) had an important influence on twentieth-century historians of mathematics, attributes, for the early period, high importance to the Latin works of Fibonacci and Jordanus. He believed that the vernacular tradition of practical arithmetic and algebra did not produce any men capable of understanding the works of these two giants. Cantor assumes this to be true for most of the 14th and 15th century.[2] When dealing with the sixteenth century, extraordinary importance is attributed to the *Arithmetica* of Diophantus (Heath, 1885; Sesiano, 1982; Rashed, 1984). The idea that algebra originated with Diophantus was fabricated by humanist mathematicians after Regiomontanus's Padua lecture of 1464 (Regiomontanus, 1537). As a consequence of their reform of mathematics, humanist writers distanced themselves from "barbaric" influences and created the myth that all mathematics, including algebra descended from the ancient Greeks (Høyrup, 1996). Later writers such as Ramus (1560; 1567); Peletier (1554); Viète (1591) and Clavius (1608) participated in a systematic program to set up sixteenth-century mathematics on Greek foundations. The late discovered *Arithmetica* of Diophantus was taken as an opportunity by Viète to restore algebra "which was spoiled and defiled by the barbarians" to a fictitious pure form. To that purpose he devised a new vocabulary of Greek terms to cover up the Arab roots of algebra "lest it should retain its filth and continue to stink in the old way" (Klein, 1968, p. 318). The reality was that, with some exceptions, ancient Greek mathematics was more foreign to European mathematics than Indian and Arabic influences which were well digested within the vernacular tradition (Heeffer, 2007).

---

[1] We follow the convention to name the abbaco or abbacus tradition after Fibonacci's Liber abbaci, spelled with double b to distinguish it from the material abbacus. It refers to the method of counting and calculating with hindu-arabic numerals.

[2] "Aber die Zeitgenossen der beiden grossen Männer waren nicht reif, deren Schriften vollständig zu verstehen, geschweige denn sie fortzubilden, und besonders für die eigentlichen Gelehrtenkreise gilt dieses harte Urtheil auch noch im XIV. Jahrhunderte, während damals italienische Kaufleute der Algebra so viel Verständniss entgegenbrachten, dass wenigstens versucht wurde, Aufgaben zu lösen, welchen die früheren Schriftsteller ohnmächtig gegenüberstanden." (Cantor, 1892)

Contemporary scholars, such as Rafaella Franci and Laura Toti Rigatelli (1979; 1985) or van der Waerden (1985) narrate the story of the history of algebra from their internal dynamics. Early European algebra, as inherited from the Arabs, recognized six types of equations. Instead of dealing with the general form of a quadratic equation $ax^2 + bx + c = 0$, the first Latin translations distinguish three cases depending on the sign of the coefficients and three cases with one or two terms missing. Each case had its own solution method. Double solutions were not recognized except for the case of two positive roots. Early abbacus masters extended the list of six to include higher degree cases most of which could be reduced to the original six. During the fourteenth century Maestro Dardi of Pisa expands this to no less than 198 cases (van Egmond, 1983; Franci, 2001)! Already from the fourteenth century abbacus masters started experimenting with irreducible cases of higher degree. False rules were given for special cases of the cubic equation (Høyrup, 2009). However, Maestro Dardi gives some examples of cubic equations with a correct solution derived from numerical examples. These histories of algebra then focus on the sixteenth-century breakthrough of Scipione del Ferro in solving the depressed cubic and the feud between Tartaglia and Cardano for publishing a general solution for the cubic equation. It then moves to symbolism introduced by Viète and the general approach to problems. With a mention of Girard, these developments culminate into the quest for the fundamental theorem of algebra moving well into the eighteenth century. All this is presented as a continuous flow of necessary logical development. Each step is the necessary next move in the logical puzzle of the history of symbolic algebra. Van der Waerden goes to great lengths of demonstrating this continuity back to the earliest Greek mathematics. In a earlier publication, van der Waerden (1988, p. 116), uses the so-called *Bloom of Thymaridas* to connect algebra with the Pythagoreans. He goes as far as to claim that "we see from this that the Pythagoreans, like the Babylonians, occupied themselves with the solution of systems of equations with more than one unknown". We have elsewhere demonstrated that these claims cannot be sustained and that these should be understood as a by "humanist education deeply inculcated prejudice that all higher intellectual culture, in particular all science, is risen from Greek soil" (Heeffer, 2009).

Jacob Klein, a student of Heidegger and interpreter of Plato, wrote a long treatise in 1936 on the number concept starting with Plato and the development of algebra from Diophantus to Viète (Klein, 1968). It became very influential for the history of mathematics after its translation into English in 1968. Klein goes even further than Van der Waerden or Franci and Rigatelli in their internalistic account of history. For Klein it is not the evolution of solution methods for solving equations which follows some logical path but the ontological transformation of the underlying concepts within an ideal Platonic realm. He restricts all other possible understandings of the emergence of symbolic algebra by formulating his research question as follows: "What transformation did a concept like that of *arithmos* have to undergo in order that a 'symbolic' calculating technique might grow out of the Diophantine tradition?" (Klein, 1968, p. 147). According to Klein it is ultimately Viète who "by means of the introduction of a general mathematical symbolism actually realizes the fundamental transformation of conceptual foundations" (Klein, 1968, p. 149). Klein places the historical move towards the use of symbols with Viète and thus ignores important contributions by the abbaco masters, by Michael Stifel (1545, 1553), Girolamo Cardano (1539, 1545) and the French algebraists Jacques Peletier (1554), Johannes Buteo (1559) and Guillaume Gosselin (1577). The new environment of symbolic representation provides the opportunity to "the ancient concept of arithmos" to "transfer into a new conceptual dimension" (Klein, 1968, p. 185). As soon as this happens, symbolic algebra is born: "A soon as 'general number' is conceived and represented in the medium of species as an 'object' in itself, that is, symbolically, the modern concept of 'number' is born" (Klein, 1968, p. 175). It is hard to understand why a philosophy like this, rooted in German idealism, where concepts realize themselves with the purpose to advance mathematics, is so appealing to modern historians looking for an explanation for the emergence of symbolic algebra.

The three different approaches to the history of algebra are exemplified by three historians of mathematics. Cantor believes in a continuous development from ancient Greek notions of number and proof to modern algebra, only obscured during the medieval period in which the old masters were not fully understood. Van der Waerden and Franci see a historic realization of the logical development from quadratic equations to cubic and higher degree ones towards a theory of the structure of equations. Klein discerns a realization of symbolic algebra in a necessary ontological transformation of the underlying number concept. All three share the idea that there is some internal necessity and logic in the historical move towards symbolic algebra. But all pass by at the fundamental historical changes that took place in the context in which medieval algebra matured: the mercantile centers of northern Italy and the French Provençe region. We shall now look at some contextual explanations and further demonstrate that the emergence of symbolic algebra cannot be understood without accounting for the socio-economical context of that time. We shall illustrate this with a specific class of bartering problems which were discussed in arithmetic and algebra books during several centuries. We do not want to present such socio-economical interpretation as 'the right one', but at least as a complementary one to the one-sided internatistic interpretation so dominant in the historiography of algebra.

3 Contextual approaches

From the 1920's, history of science began to account for contextual aspects of the society in which science develops and is practiced. As a reaction to the romantic narratives of Great Men making Great Discoveries in science communist historians of science pointed out the role of social and economic conditions in the emergence and development of science. Gary Werskey (1978) describes how Soviet historians irritated Charles Singer, the chairman of the Second International Congress in the History of Science and Technology in London in 1931, by repeatedly asking questions about socio-economical influences on the evolution of science. But this conference was a historical one making an impact on the thought of many young scholars with socialist sympathies such as Joseph Needham and Lancelot Hogben. Beginning with Boris Hessen's *The Social and Economic Roots of Newton's Principia* (1931), several papers and books were published, placing the achievements of individual scientists within the context of social superstructures. Specific histories of mathematics based on an analysis of socio-economical conditions appeared much later. Dirk Struik was a convinced Marxist who wrote a widely read *A Concise History of Mathematics* (1987). Although the book cannot be considered a Marxist analysis, his vision of mathematics as a product of culture and evolving within a dialectic process was having an impact on other historians.

Only a limited number of authors focused on the mathematical sciences during the period that symbolic algebra developed in Europe (between 1300 and 1600). Michael Wolff (1978) in a comprehensive study of the concept of impetus argues that the "new physics" of the fourteenth century developed from contemporary social thought. The scientific revolution basically was a revolution in socio-economical ideas. Drawing upon the theories of Marx and Borkenau, Richard Hadden (1994) develops the idea that practitioners of commercial arithmetic, as a consequence of their social relations, delivered the new concept of "general algebraic magnitude" to the new mechanics. Joel Kaye (1998) argues that the transformation of the model of the natural world of the Oxford and Paris scholars such as Thomas Bradwardine, John Buridan, and Nicole Oresme during the fourteenth century follows the rapid monetization of the European society. This transformation happened beyond the university and outside the culture of the book.

We would like to argue that symbolic algebra functioned together with double-entry bookkeeping as the main instruments for the determination of objective value, the basic idea of the mercantile society. The foundations for symbolic algebra were laid within the abbaco tradition. While scholars on this tradition, such as Jens Høyrup (2007) maintain that the problem solving treatises written by abbaco masters served no practical purpose whatsoever, we argue that their activities and writings delivered an essential contribution to Renaissance mercantilism in the creation of objective, computable value. According to Foucault (1966, p. 188) the essential aspect for the process of exchange in the Renaissance is the representation of value. "In order that one thing can represent another in exchange, they must both exist as bearers of value; and yet value exists only within the representation (actual or possible), that is, within the exchange or the exchangeability". The act of exchanging, i.e., the basic operation of merchant activity, both determines and represents the value of goods. To be able to exchange goods, merchants have to create a symbolic representation of the value of their goods. All merchants involved must agree about this common model to complete a successful transaction. As such, commercial trade can be considered a model-based activity. Given the current global financial market and the universal commensurability of money we pass over the common symbolic representation as an essential aspect of trade. However, during the early Renaissance, the value of money depended on the coinage, viz. the precious metals contained in the coins which differed between cities, and varied in time. As the actions and reciprocal relations of merchants, such as exchange, allegation of metals and bookkeeping became the basis for the symbolic and abstract function of money, so did the operations and the act of equating polynomials lead to the abstract concept of the symbolic equation. Both processes are model-based and use the symbolism as the model. Therefore, we have to understand the emergence of symbolic algebra within the same social context as the emergence of double-entry bookkeeping.

Now consider the following statement: The emergence of double-entry bookkeeping by the end of the fifteenth century was a consequence of the transformation from the traveling to the sedentary merchant, primarily in the wool trade situated in Italy and Flanders (de Roover, 1948; van Egmond, 1976). Given the vast body of evidence from Renaissance economic history and the evident causal relationship, not many will contest the relevance of merchant activities on the emergence of bookkeeping. What about the mitigated statement: "The emergence of symbolic algebra in the sixteenth century is to be situated and understood within the socio-economic context of mercantilism"?. Philosophers of mathematics who believe in an internal dynamics of mathematics will not accept decisive social influences as an explanation for the emergence of something as fundamental as symbolic algebra. At best, they will accept social factors in the acceleration or impediment of what they consider to be a necessary step in the development of mathematics. Also it seems difficult to pinpoint direct causal factors within economic history for explaining new developments in mathematics. However, the relationship between bookkeeping and symbolic algebra is quite remarkable. Many authors who have published about bookkeeping also wrote on algebra. The most notorious example is Pacioli's *Summa*, which deals with algebra as well as bookkeeping, and the book had an important influence in both domains. But there are several more coincidences during the sixteenth century. Grammateus (1518) gives an early treatment of algebra together with bookkeeping. The Flemish reckoning master Mennher published books on both subjects including one treating both in the same volume (Mennher, 1565). So did Petri in Dutch (Petri, 1583). Simon Stevin wrote an influential book on algebra (Stevin, 1585) and was a practicing bookkeeper who wrote a manual on the subject (Stevin, 1608). In Antwerp, Mellema published a book on algebra (Mellema, 1586) as well as on bookkeeping (Mellema, 1590). While there is no direct relationship between algebra and bookkeeping, the teaching of the subjects and the books published often addressed the same social groups. Children of merchants were sent to reckoning schools (in Flanders and Germany) or abbacus schools (in Italy) where they learned the skills useful for trade and commerce. There is probably no need for algebra in performing bookkeeping operations but for complex bartering operations or the calculation of compound interest, basic knowledge of arithmetic was mandatory and knowledge of algebra was very useful.

4 Case study: Bartering with cash values

In an interesting article in the *Journal of the British Society for the History of Mathematics*, John Mason (2007) expresses his surprise at the solution method adopted for bartering problems which involve cash. He cites a problem from Piero della Francesca in a translation by Judith Field (2005, p. 17):

> Two men want to barter. One has cloth, the other wool. The piece of cloth is worth 15 ducats. He puts it up for barter at 20 and $1/3$ in ready money. A cento of wool is worth 7 ducats. What price for barter so that neither is cheated?

Mason originally expected the solution to be based on the proportion of the barter value to the original value with the barter value being "either 20 + 20/3 ducats or to 20 + 15/3 ducats, depending on which value the $1/3$ is intended to act upon" However Piero's solution appears to be different (Mason, 2007, p. 161):

> This computation intrigued me because I was astonished at the sequence of calculations: first reduce by the ready money paid (as a fraction of the barter price), and only then compare barter prices. It seemed to me that in a modern economy it would be more natural to carry out one of the calculations I considered, since the ready money to be paid is a cash value, and the bartering inflation refers to the noncash-traded amounts.

Thus Piero subtracts one third of 20 from 20, which leaves 13 $1/3$ and the same value from 15 which becomes 8 $1/3$. The proportion of these two values is hence the fair barter profit to be applied by both parties. Though Mason lists several other examples which follow the same solution method, he does not provide an explanation why this particular method is adopted in abbaco treatises and in later printed books. Given that this way of calculating was in use for over two centuries, not only in Italy but in several European countries, this particular bartering practice needs an explanation. We shall demonstrate that his astonishment is based on a wrong interpretation and even more so, a wrong translation of the original problem. We shall provide an explanation by placing these early bartering problems within the specific context of Medieval Italian merchant practices.

The original problem by Piero, in Gino Arrighi's transcription from the manuscript, is formulated as follows (f. 8r; Arrighi, 1970, p. 49):

> Sono doi che voglano baractare, l'uno à panno e l'altro à lana. La peçça del panno vale 15 ducati et mectela a baracto 20 et sì ne vole $1/3$ de contanti; et il cento de la lana vale 7 ducati a contanti. Che la dèi mectere a baracto a ciò che nisuno non sia ingannato?

A literal translation of the medieval Italian would be as follows:[3]

> There are two [men] that want to barter. One has cloth, the other has wool. The piece of cloth is worth 15 ducats. And he puts this to barter [at] 20 and of this he wants $1/3$ in cash. And a hundred of wool is worth 7 ducats in cash. What shall they put for barter so that not one of them is being cheated?

Formulated this way, there is little room for doubt. The one with the cloth wants 20 ducats per piece of cloth, *of which one third in cash*. Obviously then, a third of the value refers to the barter value of 20. The amount of cash per piece is thus 20/3. To know the barter value of the cloth without the cash one has to subtract the cash from it, being 13 $1/3$. That Mason wants to add one third of the value rather than subtracting it stems from the wrong translation of "et sì ne vole $1/3$ de contanti".

Is this interpretation the correct one for all bartering problems of this type in abbaco treatises? Let us look for further clues. Mason provides pointers to several abbaco treatises in which bartering problem appear with a cash value. The earliest he discusses are problems 33, 86, and 87 of Paolo Dagomari's *Trattato d'aritmetica*, written in 1339. He describes problem 86 as a problem which "involves grain to be bartered at 15s but valued at 12, with one-third in ready money, in exchange for *orzo* (?) at 10s.". The word

---

[3] My translation. For a discussion on the translation of abbaco texts, cf. a recent critical edition of a fifteenth-century treatise on algebra (Heeffer, 2008, p. 132)

orzo should pose no problems as it is the modern Italian word for barley. In our translation:[4]

> There are two that want to barter together. The one has grain and the other has barley. And the one with the grain which is valued at 12 s. puts it in barter at 15 s. per bushel. And he wants from the one with the barley one third of the value in cash. And from what remains he will get barley. And a bushel of barley values 10 s. Asked is what they arrive at in this barter so that none is left cheated.

Here also, the meaning of the problem is different from the one paraphrased by Mason. It is not the person with the grain who puts in the cash, but the other one. Furthermore, the enunciation clearly specifies that the second person should deliver one third of the value in cash and the rest in barley and this conforms with our interpretation.

4.1 First occurrences of bartering with cash

Was Paolo the first to deal with cash values in bartering problems? We checked all available transcriptions of abbaco treatises before Paolo's *Trattato*. The earliest one is probably the *Columbia algorismus* (Columbia, X 511, A 13) published by Vogel (1977). Vogel himself dated the manuscript in the second half of the 14th century. However, a recent study of the coin list contained in the manuscript is dated between 1278 and 1284, which makes it the earliest extant treatise within the abbaco tradition (Travaini, 2003, pp. 88-92). Høyrup suspects it "likely to be a copy of a still earlier treatise" (Høyrup, 2007, p. 31). It contains two barter problems (19 and 20) but none involves money. The anonymous *Livero del l'abbecho* is dated c. 1289–1290 and has also two bartering problems without money (Arrighi, 1989, p. 24, 28). The *Tractatus Algorismi* by Jacopo da Firenze is extant in an earliest version of 1307. It is the subject of a recent comprehensive study of the abbaco tradition by Jens Høyrup (2007). However this extensive treatise does not contain any bartering problems. The next available transcription is the *Liber habaci*, dated by van Egmond (1980) to 1310, and is the first to involve cash in a bartering transaction. The enunciation of the single bartering problem is more elaborate and functions as a prototype for later reformulations by Paolo and Piero:[5]

---

[4]From Arrighi's transcription (Arrighi, 1964, p. 75): "E' xono due che barattano insieme, l'uno àe grano e Il'altro àe orzo; e quello che à grano gli mette in baratto lo staio del grano 15 s., che vale 12 s., e vuole il terzo da quello dell'orzo di ciò che monta il suo grano di chontantj; e dell'avanzo se ne togle orzo. Ello staio dell'orzo vale 10 s., adornando quanto glele chonterà in questo baratto acciò che no' rrimangha inghannato".

[5]From Liber habaci, Biblioteca Nazionale Centrale Firenze, Magl. Cl. XI, 88, transcription by Arrighi (1987, p. 147): "Sono due merchatanti che volglono barattare insieme, l'uno si à lana e l'altro si à pannj; dice quellj ch'à lla lana a quellj del pannj: che vuo'

There are two merchants who want to barter together. The one has wool and the other has cloth. The one with the wool tells the one with the cloth: "how much do you want for the *channa* of your cloth". And he says: "I want 8 lb. (and he knows well that it values not more than 6 lb.) and I want one quarter in cash and I want three quarters in wool. And the wool is valued at 20 lb. per hundred. Asked is what suits him to sell the wool per hundred so that he is not being cheated.

We find here all the elements of the later bartering problems. The problem clearly specifies that one party will deliver one quarter of the value in cash and three quarters in merchandize. The reference to a fair deal becomes a standard formulation in abbaco bartering problems. The solution recipe is the standard formula adopted in later treatises as discussed by Mason:[6]

> You shall do as such, one quarter is asked in cash, say as such: one quarter of eight is 2. The rest until eight is 6. From 2 until 6 is 4, therefore say as such: for every 4 lb. I get 6 lb., how much do I get for 20 lb.? Multiply 20 lb. against 6 lb. this makes 120 lb. Divide by 4 and 30 lb. results from it. This is how much it suits him to get per hundred for this wool.

We have now found an adequate interpretation for the subtraction of the cash value from the barter price, but why is this cash value also subtracted from the original value? This example from the *Liber habaci* already gives us an insight. Obviously, if one takes into account the barter value minus the cash value (here 6 lb.) something also has to be done with the original value of the merchandize (also 6 lb.). In this example these values are the same and there would be no profit ratio. However, adapting Mason's original reasoning to the new interpretation, one could still compare the total barter value (here 8 lb.) with the original value (6 lb.) and use this as a profit ratio. Why is it not done this way?

4.2 Early Italian merchant practices

To answer that question we must look at Italian merchant practices at the beginning of the fourteenth century. One important breakthrough took place around that time: the introduction of double-entry bookkeeping.

---

tu della channa del panno? E que' dice: io ne volglo lb. viij (e sa bene che non vale più di lb. vj) e volglo il quarto i' d. chontanti e tre quarti volglo in lana. El centinaio della lana vale lb. xx, adomando che lgli chonviene vendere il centinaio di questa lana acciò che non sia inghannato". A channa is a unit of length of about 2 m.

[6]Ibid: "De' chosì fare. E' domanda il quarto in danari, diray chosì: il quarto d'otto si è ij. Insino inn otto si à vj, da ij insino vj si iiij or diray chosì: ongnj iiij lb. mi mette lb. vj, che mi metterà lb. xx? Multipricha lb. xx via lb. vj farà lb. Cxx, dividi per iiij ne viene xxx lb.: chotanto gli chonviene mettere il centinaio di questa lana".

Records of stewards of authorities of Genoa in 1278 show no trace of this kind of bookkeeping while by 1340 a complete system of double-entry bookkeeping was established (Littleton, 1927, p. 147). While archival evidence suggests the emergence of bookkeeping practices during the course of the thirteenth century, the earliest extant evidence of full double-entry bookkeeping is the Farolfi ledger of 1299–1300 (Lee, 1977). So the appearance of cash in bartering problems during the first decades of the fourteenth century coincides with the emergence of double-entry bookkeeping practices. Bartering was the dominant practice for traveling merchants during the Middle Ages. When medieval Europe moved to a monetary economy in the thirteenth century, sedentary merchants depended on bookkeeping to oversee multiple simultaneous transactions financed by bank loans. While standard bartering required no elaborate administration, double-entry bookkeeping supported more complex bartering operations involving cash and time. Calculating practices taught in *bottega d'abbaco*, supported the new economy in the same way as double-entry bookkeeping did. If we want to understand these problems we should therefore look at bookkeeping practices.

As is well known, the first printed text on double-entry bookkeeping is Pacioli's *Particularis de Computis Et Scripturis*, treatise XI of distinction nine of his *Summa de arithmetica et geometria* of 1494. Mason cites from the *Scripturis* but oddly not from the chapter 20 on bartering. Pacioli was well aware about the old bartering practices. His until recently unpublished Perugia manuscript (Vat. Lat. 3129, 1478) contains a chapter on bartering with no less than 56 problems (folios 61^r–83^v). Many of them involve cash (Calzoni and Cavazzoni, 1996). In the *Scripturis* he writes "Bartering is commonly of three kinds: Simple, Complex, and Time" (*Semplice, Composta, a Tempo*) and he explains how to account for bartering in the Journal and Ledger (Crivelli, 1924, p. 46):

> After you have so described it, you can then reduce it to cash value, and as you wish to see the value in cash of such and such goods you will make out the entry in the Memorandum in whatever kind of money you desire; as it does not matter, providing that the bookkeeper afterwards transfers the entry to the Journal and Ledger and reduces the amount to the standard money which you have adopted.

Our bartering problems involving cash are thus of the complex type and Pacioli provides an example of how to note down the value of bartered merchandise for a transaction which involves one third in cash. In Pacioli's terminology *Per* stands for debit and *A* for credit (Crivelli, 1924, p. 47):

> *Per Bellidi* ginger in bulk or packages. *A* sugar of such and such a kind, so many packages, weighing so much. Received ginger from so-and-so in exchange for sugar carried out in this manner: viz., I valued

the sugar at 24 ducats per hundred, on condition that I should receive 1/3 in cash, and the ginger to be valued at so many ducats per hundred, for which ginger I should give so many loaves of sugar, weighing so much, which if paid for in cash are worth 20 ducats per hundred, and for said ginger he received sugar, so many loaves, each valued at L[ire] S[oldi] G[rossi] P[icioli]

Unfortunately Pacioli gives no numerical entries but explains that one should debit the cash (you receive) and credit the sugar (you barter). Furthermore "that which is more in the cash entry will nevertheless be missing per contra in the sugar, and this you are to correct". So let us reconstruct the bookkeeping transactions for the original example by Piero given the balance sheet equation: Assets = Liabilities + Owners Equity, and using Pacioli's [Debit // Credit] notation system:[7]

Assets	Liabilities	Owners Equity

1) Spend 1/3 cash of barter price

[0 // 20/3]   [20/3 // 0]

2) Deliver products for barter from stock

[15 // 0]

3) Receive barter goods at barter value

[0 // 20]

4) Book profit

[0 // 20 − 15]

One third of the barter value is paid in ready money and therefore credited from cash assets and debited from OE. The products we deliver have a booking value of 15 (multiplied by the number of items) while we receive goods valued at 20. The difference has to be booked as profit to maintain the balance and therefore we credit OE with the difference, being 20 − 15. We now can see that the profit or difference between the booked value and the barter value has to be the same as the difference between the two values

---

[7] As the balance sheet equation was introduced after Pacioli this may seem an anachronism. It is here only shown to demonstrate the necessity of step 4.

used to determine the fair profit ratio. Thus the calculation of the barter value of the second party, x depends on the ratio:

$$\frac{20 - \frac{1}{3}(20)}{15 - \frac{1}{3}(20)} = \frac{x}{7}$$

Subtracting the two values $20 - \frac{1}{3}(20)$ and $15 - \frac{1}{3}(20)$ results in the profit 5.

That a seemingly basic problem from the abbaco tradition, which follows practices that were in use for over two centuries, gives rise to a feeling of astonishment for modern scholars on the history of mathematics is rather interesting. That we have to base ourselves on the socio-economical context of mercantilism to understand the solution of the problem is even more so. The case demonstrates that starting from modern conceptions and looking for corresponding ones in a historical context is often not the best way to study history. Ideas, methods and practices, even mathematical ones, are best understood in their historical socio-economical context.

4.3 Concluding remarks

The earliest written evidence of double-entry bookkeeping is the Farolfi ledger of 1299–1300. The earliest extant vernacular text dealing with algebra was written in 1307. Both algebra and double-entry bookkeeping were practiced throughout the fourteenth and fifteenth centuries in the mercantile centers of Northern Italy. The first appearance of these two disciplines in print was in the very same book, the *Summa* by Luca Pacioli (1494). Why was bookkeeping treated in the same book with arithmetic and algebra? Because they both were important instruments in the establishing objective and exact value, the basic principle of reciprocal relations of exchange in a mercantile society. There has been speculation about the purpose and intended audience of a book like the *Summa*. While Pacioli himself received an abbaco education and he taught some years to sons of merchants in Venice, he is often wrongly considered an abbacist (e.g., Biagioli, 1989). In fact, he enjoyed the social status of a well-paid university professor. Between 1477 and 1514, he taught mathematics at the universities of Perugia, Zadar (Croatia), Florence, Pisa, Naples and Rome (Taylor, 1942). This Franciscan friar and university professor saw a way to bring his lengthy treatise in Tuscan vernacular to a larger public by means of book printing. Later he would publish a Latin book on Euclidean geometry but as an educator he recognized the real needs of the mercantile society. His *Summa* literally brings together all important aspects of knowledge in such a mercantile society, including algebra and double-entry bookkeeping. In a recent study on the target audience (Sangster et al., 2008, p. 129), Sangster is

> ... led to the conclusion that the bookkeeping treatise was not only intended to be read and used by merchants and their sons, it was

designed specifically for them. Further analysis of the content and sequencing of *Summa Arithmetica* indicated that the entire book was written primarily as a reference text for merchants and as a school text for their sons. It was sourced mainly from *abbaco* texts and mirrored much of the curriculum of the *abbaco* schools attended by the sons of merchants; and extended it to include all extant material known to Pacioli that was of direct relevance to merchants. No abbaco school or tutor would previously have had access to such a wide range of relevant material in a single source.

From the sixteenth century, under the influence of the humanist program to reform mathematics, algebra changed considerably and by the end of the sixteenth century developed into a symbolic algebra. On the other hand, double-entry bookkeeping did not and is in use today in the smallest coffee shops to the largest multi-national enterprizes, in a way which is not fundamentally different from Pacioli's description. This is remarkable as the structure of the ledgers is designed in a way to avoid negative numbers (Peters and Emery, 1978, wrongly critized by Scorgie, 1989). The balance sheet formula used above can equally be expressed as Assets−Liabilities−Owners Equity $= 0$, in accordance with our common way of writing down equations. However, this involves the use of negative quantities, a concept which was gradually introduced only from the sixteenth century onwards. However strange this may sound, our modern way of doing bookkeeping thus conserves the medieval concept of number.

As far as we know it was Maestro Antonio de' Mazzinghi who was the first to apply the rules of algebra to bartering problems in the 1380's. In his *Trattato di fioretti* he solves a simple problem of two men bartering wool against cloth (Arrighi, 1967, pp. 31-2). The wool is worth 20 and the barter value is 22. The cloth is worth 6 and bartered with the same profit margin adjusted by 10%. Instead of doing the numerical calculation de' Mazzinghi solves this by taking x for the barter profit so that $\frac{1}{1x} + \frac{1}{1x+1}$ equals 10%. This leads him to the equation

$$\frac{2x+1}{x^2+x} = \frac{1}{10}$$

By solving the quadratic equation he arrives at a barter price for the wool as $\sqrt{401} - 1$ *fiorini* against the price for the cloth at $\sqrt{43\frac{6689}{10000}} + \frac{33}{100}$ *fiorini*.

We can find surds as values for goods in almost every abbaco treatise. Høyrup (2009, p. 51) appropriately remarks that "Being satisfied with exactly expressed but irrational solutions remained the habit of abbacus algebra". In contrast with geometry treatises which served the purpose of practical surveying and construction and had its values approximated. You do not find any approximations within the context of abbaco algebra. This leads him to the conclusion that "abbacus algebra, at least beyond the first degree, must in some sense have been a purely theoretical discipline without intended practical application". However, as we see it, in their perseverance on using exact values for merchant type problems, abbacus masters established the objective true value of goods within a transaction. If this had to be expressed in surds, so be it. The underlying idea was that there always exists a just and true value in exchange. Earlier in medieval society the marketplace was already recognized as a guide to the determination of value. But this value was a fuzzy concept, a value subject to many approximations in calculating the exchange rates of currencies and unit conversions which could be different from city to city.

The earliest extant abbaco text dealing with algebra, Jacoba da Firenzi's *Tractatus Algorismi* of 1307, includes a peculiar discounting problem (Høyrup, 2007, pp. 252-3):

> A merchant shall have from another *libre* 200 within two months and a half from now. This merchant says, give me this money today, and I discount your money at the rate of *denarij* ij [i.e., 2] per *libra* a month. Say me, how much shall he give him in advance for the said *libre* 200.

The problem is followed by an approximate solution based on iterative subtractions. Høyrup (2007, p. 70, note 176) reports a similar approximation in the fifteenth-century *Libro di conti e mercantanzie* (Gregori and Grugnetti, 1998, p. 95). Expressed in modern symbolism this amounts to the development of $\frac{1}{1+p}$ as $1 - p(1 - p(1 - p(1 - p(1 - p))))$. One could see these two occurrences as counter examples in the development towards exact value within the abbaco tradition. However, it is important to note that such old merchant calculations are quite rare and that they became gradually replaced by algebra. It is precisely within the algebraic context that we see no approximate solutions. One of the major contributions of the maestri d'abbaco is that they have shown that practically all merchant problems can be approached by algebra. We find a good illustration of this in cumulative interest problems which could be solved by an iterative procedure similar to the one described. However, we see that these problems were successfully solved by algebra even if they lead to fifth-degree equations as in Piero della Francesca's *Trattato d'abacho* (Arrighi, 1970, pp 421–2).

5 Conclusion

Abbaco arithmetic and algebra as well as double-entry bookkeeping gave support to the idea of a value which can be determined in principle as the just value, not only in the mathematical sense but also in the moral sense. The frequent references to the *fair* barter value demonstrate the moral obligation to account for exact values. The importance of dealing with exact values should not be underestimated. Margolis's barrier theory places the concept of an exact quantifiable number for probabilities as the breakthrough for the probability theory of Pascal and Fermat (Margolis, 1993). All intuitions for a probability theory were present for centuries, but the habit of mind was to perceive probability as a result of a bargaining process, like a fair price for the risks involved. Probability theory was made possible by attaching a single quantifiable number to the concept of probability which was not considered to be countable. Once this missing concept was introduced of usefully attaching numbers to comparative values even if there is nothing immediately to count, a theory of probability could be established. The idea of an exact value, which could be expressed within the language of algebra and maintained its exactness through the operations of algebra, became the basic concept of mercantilism.

Bibliography

Arrighi, G., editor (1964). *Paolo dell'Abaco, Trattato d'Aritmetica. Secondo la lezione del Codice Magliabechiano XI,86 della Biblioteca Nazionale di Firenze.* Domus Galileana, Pisa.

Arrighi, G., editor (1967). *Antonio de' Mazzinghi. Trattato di Fioretti, secondo la lezione del codice L.IV.21 (sec. XV) della Biblioteca degli Intronati di Siena.* Domus Galileana, Pisa.

Arrighi, G., editor (1970). *Piero della Francesca. Trattato d'abaco: dal codice Ashburnhamiano 280 (359*–291*) della Bilioteca Medicea Laurenziana di Firenze.* Domus Galileana, Pisa.

Arrighi, G., editor (1987). *Paolo Gherardi. Opera mathematica: Libro di ragioni – Liber habaci. Codici Magliabechiani Classe XI, nn. 87 e 88 (sec. XIV) della Biblioteca Nazionale di Firenze.* Pacini-Fazzi, Lucca.

Arrighi, G. (1989). Maestro umbro. (sec. xiii) livero de l'abbecho. (cod. 2404 della biblioteca riccardiana di firenze). *Bollettino della Deputazione di Storia Patria per l'Umbria*, 86:5–140.

Biagioli, M. (1989). The social status of Italian mathematicians 1450–1600. *History of Science*, 27:41–95.

Boncompagni, B. (1857). *Scritti di Leonardo Pisano, matematico del secolo decimoterzo (2 vols.).* Tipografia delle Scienze Matematiche e Fisiche, Rome. I. Liber Abbaci di Leonardo Pisano, II. Leonardi Pisani Practica geometriae ed opuscoli.

Buteo, J. (1559). *Logistica*. Gulielmum Rovillium, Lyon.

Calzoni, G. and Cavazzoni, G., editors (1996). *Tractatus Mathematicus ad Discipulos Perusinos*. Città di Castello, Perugia.

Cantor, M. (1880). *Vorlesungen über Geschichte der Mathematik. Volume I: Von den ältesten Zeiten bis zum Jahre 1200 n. Chr.* Teubner, Leipzig.

Cantor, M. (1892). *Vorlesungen über Geschichte der Mathematik. Volume II: Von 1200–1668*. Teubner, Leipzig.

Cardano, G. (1539). *Practica arithmetice & mensurandi singularis. In qua que preter alias continentur, versa pagina demonstrabit*. Bernardini Calusci, Milan.

Cardano, G. (1545). *Artis magnæ, sive, De regulis algebraicis lib. unus: qui & totius operis de arithmetica, quod opus perfectum inscripsit, est in ordine decimus*. Johannes Petreius, Nürnberg.

Clavius, C. (1608). *Algebra Christophori Clavii*. Zanetti, Rome.

Crivelli, P. (1924). *Double-entry book-keeping, Fra Luca Pacioli*. Institute of Book-keepers, London.

de Roover, R. (1948). *Money, banking and credit in mediaeval Bruges: Italian merchant-bankers, Lombards and money-changers: a study in the origins of banking*. Mediaeval Adademy of America, Cambridge, MA.

Field, J. F. (2005). *Piero della Francesca: a mathematician's art*. Yale University Press, London.

Foucault, M. (1966). *Les mots et les choses. Une archéologie des sciences humaines*. Gallimard, Paris. Page numbers refer to the English translation (Foucault, 1971).

Foucault, M. (1971). *The order of things: an archaeology of the human sciences*. Pantheon Books, New York.

Franci, R. (2001). *Maestro Dardi (sec. XIV). Aliabraa argibra. Dal manoscritto I.VII.17 della Biblioteca Comunale di Siena*, volume 26 of *Quaderni del Centro Studi della Matematica Medioevale*. Università di Siena, Siena.

Franci, R. and Rigatelli, L. T. (1979). *Storia della teoria delle equazioni algebriche*. Mursia, Milan.

Franci, R. and Rigatelli, L. T. (1985). Towards a history of algebra from Leonardo of Pisa to Luca Pacioli. *Janus*, 72(1–3):17–82.

Gosselin, G. (1577). *De arte magna, seu de occulta parte numerorum, quae & algebra, & almucabala vulgo dicitur, libri qvatvor. Libri Qvatvor. In quibus explicantur aequationes Diophanti, regulae quantitatis simplicis, et quantitatis surdae*. Aegidium Beys, Paris.

Grammateus, H. (1518). *Ayn new kunstlich Buech welches gar gewiss vnd behend lernet nach der gemainen Regel Detre, welschen practic, Reglen falsi vñ etlichäe Regeln Cosse*. Johannem Stüchs, Nürnberg.

Gregori, S. and Grugnetti, L., editors (1998). *Libro di conti e mercatanzie. Anonimo (sec. XV)*. Università degli Studi di Parma, Facoltà di Scienze Matematiche, Fisiche e Naturali, Dipartimento di Matematica, Parma.

Hadden, R. (1994). *On the shoulders of merchants. Exchange and the mathematical conception of nature in early modern Europe*. State University of New York Press, Albany.

Heath, T. L. (1885). *Diophantus of Alexandria: A study in the history of Greek algebra*. Cambridge University Press, Cambridge.

Heeffer, A. (2007). The tacit appropriation of Hindu algebra in renaissance practical arithmetic. *Gaṇita Bhārāti*, 29(1–2):1–60.

Heeffer, A. (2008). Text production, reproduction and appropriation within the abbaco tradition: A case study. *Sources and Commentaries in Exact Sciences*, 9:101–145.

Heeffer, A. (2009). The reception of ancient Indian mathematics by western historians. In Yadav, B., editor, *Ancient Indian leaps in the advent of mathematics*, pages 135–152, Basel. Birkhäuser.

Hessen, B. (1931). The social and economic roots of Newton's principia. In Bukharin, N. I., editor, *Science at the crossroads*, pages 151–212, London. Frank Cass and Co.

Høyrup, J. (1996). The formation of a myth: Greek mathematics – our mathematics. In Goldstein, C., Gray, J., and Ritter, J., editors, *L'Europe mathématique. Mathematical Europe*, pages 103–119, Paris. Éditions de la Maison des sciences de l'homme.

Høyrup, J. (2007). *Jacopo da Firenze's* Tractatus Algorismi *and early Italian abbacus culture*, volume 34 of *Science Networks. Historical Studies*. Birkhäuser, Basel.

Høyrup, J. (2009). What did the abbacus teachers aim at when they (sometimes) ended up doing mathematics? An investigation of the incentives and norms of a distinct mathematical practice. In Van Kerkhove,

B., editor, *New perspectives on mathematical practices. Essays in philosophy and history of mathematics*, pages 47–75, Singapore. World Scientific Publishers.

Kaye, J. (1998). *Economy and nature in the fourteenth century: Money, market exchange, and the emergence of scientific thought*. Cambridge University Press, Cambridge.

Klein, J. (1968). *Greek mathematical thought and the origin of algebra*. MIT Press, Cambridge, MA. Translated by E. Brann.

Lee, G. A. (1977). The coming of age of double entry: The Giovanni Farolfi ledger of 1299–1300. *Accounting Historians Journal*, 4(2):79–95.

Littleton, A. (1927). The antecedents of double-entry. *Accounting Review*, 2(2):140–149.

Margolis, H. (1993). *Paradigms & barriers: How habits of mind govern scientific beliefs*. University of Chicago Press, Chicago, IL.

Mason, J. (2007). Bartering problems in arithmetic books 1450–1890. *British Society for the History of Mathematics Bulletin*, 22(3):1–24.

Mellema, E. (1586). *Second Volume de l'Arithmetique*. Gillis van den Rade, Antwerp.

Mellema, E. (1590). *Boeckhouder na de conste van Italien*. Gielis van den Rade, Amsterdam.

Mennher, V. (1565). *Practicque pour brievement apprendre à ciffrer & tenir livre de comptes auec la regle de Coss & get'ometrie*. Gillis Coppens van Diest, Antwerpen.

Pacioli, L. (1494). *Summa de arithmetica geometria proportioni: et proportionalita. Continetia de tutta lopera*. Paganino de Paganini, Venice.

Peletier, J. (1554). *L'algèbre de Jaques Peletier du Mans, départie en 2 livres....* J. de Tournes, Lyon.

Peters, R. M. and Emery, D. R. (1978). The role of negative numbers in the development of double entry bookkeeping. *Journal of Accounting Research*, 16(2):424–426.

Petri, N. (1583). *Practicque, om te leeren rekenen, cijpheren ende boeckhouwen, met die regel coss ende geometrie seer profijtelijcken voor alle coopluyden*. Cornelis Claesz, Amsterdam.

Ramus, P. (1560). *Algebra*. Andreas Wechel, Paris.

Ramus, P. (1567). *Scholae mathematicae*. Andreas Wechel, Paris.

Rashed, R. (1984). *Diophante. Les Arithmétiques tome III (livre IV) et tome IV (livres V, VI, VI)*. Les Belles Lettres, Paris.

Regiomontanus, J. (1537). *Alfraganus, Continentur in hoc libro. Rvdimenta astronomica alfragani. Item Albategni vs alstronomvs peritissimvs de motv stellarvm, ex obseruationibus tum proprijs, tum Ptolomaei, omnia cu demonstratiōibus Geometricis & Additionibus Ioannis de Regiomonte. Item Oratio introductoria in omnes scientias Mathematicas Ioannis de Regiomonte, Patauij habita, cum Alfraganum publice praelegeret. Eivsdem utilissima introductio in elementa Euclidis*. Nürnberg. Facsimile published in Schmeidler (1972, p. 43–53).

Sangster, A., Stoner, G., and McCarthy, P. (2008). The market for Luca Pacioli's *Summa Arithmetica*. *Accounting Historians Journal*, 111–134(1):35.

Schmeidler, F., editor (1972). *Johannis Regiomontani Opera collectanea, "Oratio Iohannis de Monteregio, habita in Patavii in praelectione Alfragani"*. Otto Zeller Verlag, Osnabrück.

Scorgie, M. E. (1989). The role of negative numbers in the development of double entry bookkeeping: A comment. *Journal of Accounting Research*, 27(2):316–318.

Sesiano, J. (1982). *Books IV to VII of Diophantus' Arithmetica in the Arabic translation attributed to Qustā ibn Lūqā*. Springer Verlag, Heidelberg.

Sigler, L. (2002). *Fibonacci's Liber Abaci. A Translation into Modern English of Leonardo Pisano's Book of Calculation*. Springer, Heidelberg.

Stevin, S. (1585). *L'Arithmetique*. Christophe Plantin, Leyden.

Stevin, S. (1608). *Livre de compte de prince à la manière d'Italie: en domaine et finance extraordinaire*. Jan Paedts Jacobsz, Leyden.

Stifel, M. (1545). *Arithmetica Integra*. Petreius, Nürnberg.

Stifel, M. (1553). *Die Coss Christoffe Ludolffs mit schönen Exempeln der Coss*. Alexandrum Lutomyslensem, Königsberg.

Struik, D. (1987). *A concise history of mathematics*. Dover, New York, revised edition.

Taylor, E. R. (1942). *No royal road. Luca Pacioli and his times.* University of Carolina Press, Chapel Hill.

Travaini, L. (2003). *Monete, mercanti e matematica. Le monete medievali nei trattati di aritmetica e nei libri di mercatura.* Jouvence, Rome.

van der Waerden, B. L. (1985). *A history of algebra from al-Khwārizmī to Emmy Noether.* Springer, Heidelberg.

van der Waerden, B. L. (1988). *Science awakening.* Kluwer, Dordrecht, 5th edition.

van Egmond, W. (1976). *The commercial revolution and the beginnings of western mathematics in renaissance Florence, 1300–1500.* PhD thesis, Indiana University.

van Egmond, W. (1980). *Practical mathematics in the Italian renaissance: A catalog of Italian abacus manuscripts and printed books to 1600*, volume 4 of *Monografia*. Istituto e Museo di Storia della Scienza, Firenze.

van Egmond, W. (1983). The algebra of Master Dardi of Pisa. *Historia Mathematica*, 10:399–421.

Viète, F. (1591). *In artem analyticam isagoge. Seorsim excussa ab Opere restituae mathematicae analyseos, seu algebra nova.* J. Mettayer, Tournon.

Vogel, K. (1977). *Ein italienisches Rechenbuch aus dem 14. Jahrhundert (Columbia X 511 A13)*, volume 33 of *Veröffentlichungen des Deutschen Museums für die Geschichte der Wissenschaften und der Technik. Reihe C: Quellentexte und Übersetzungen.* Deutsches Museum.

Werskey, G. (1978). *The visible college: A collective biography of British scientists and socialists of the 1930s.* Allen Lane, London. Reprinted 1988.

Wolff, M. (1978). *Geschichte der Impetustheorie: Untersuchungen zum Ursprung der klassischen Mechanik.* Suhrkamp, Frankfurt a.M.

François, K., Löwe, B., Müller, T., Van Kerkhove, B., editors,
Foundations of the Formal Sciences VII
Bringing together Philosophy and Sociology of Science

Economic calculation. Frameworks and performances

HERBERT KALTHOFF

Johannes Gutenberg-Universität Mainz, Institut für Soziologie, Colonel-Kleinmann-Weg 2, 55099 Mainz, Germany
E-mail: `Herbert.Kalthoff@uni-mainz.de`

1 Introduction

Within the scope of recent studies in economic sociology, Social Studies of Finance have emerged which implement the methodological research strategies and theory perspectives of *Science and Technology Studies* for the exploration of global financial markets and banking. Thus sociological knowledge, which was developed in the course of research on the interactive and technical emergence of (scientific) facts in the area of *Science and Technology Studies* (STS), is transferred to the observation of economic processes. The social studies of finance are by no means homogeneously structured; rather they make reference to various approaches within STS as well as theories of practice. This paper investigates the central role of economic knowledge, the performativity of economic representations, and the local practices that are imbedded in these forms of knowledge (e.g., Callon, 1998; Beunza and Stark, 2004; Knorr and Brüger, 2002; MacKenzie, 2006; Kalthoff et al., 2000). Against this background, a detailed discussion will be offered with regard to economic calculation and its technically framed infrastructure as exemplified by risk management in the banking sector.

Whenever large international banks grant loans to companies, they are faced with the problem of having to discern and carefully consider whether the loan can be paid back or not. In order to obtain an appraisal about the future solvency of companies, bank staff reviews their economic performance and financial standing from various angles. Firstly, they transform documents (e.g., balance sheets, profit and loss accounts) they have received from their clients and other institutions (e.g., business consultant firms) into

Received by the editors: 6 April 2009; 20 March 2011; 18 April 2011.
Accepted for publication: 18 April 2011.

their own model of economic representation; this transformation serves as a precondition for calculating a multiplicity of economic ratios. Secondly, they personally call on their (potential) clients in order to gain an impression of the company's current state and the prevailing atmosphere. They try, as one corporate banker explains, "to get an idea how the process is running [...] how the production facilities are organized, what the ambience is like, and how the workers move around". It is these direct observations of economic reality that allow "*sinnliche Gewißheit*" (Hegel, 1807, Kapitel I): the economic life-world is just as bank staff perceives it. Thirdly, they prepare a credit proposal that summarizes the results to date: purpose, maturity, term, and composition of the loan and a description of the potentially financed object, possible collaterals and internal evaluation (ratings), the company's credit standing and its economic-financial situation, various information about the debtor (address, duration of the client-bank relationship, etc.), as well as handwritten comments by the signing bank staff member vested with the relevant credit authority. This credit proposal will then be negotiated within the subsidiaries of large international banks and among the subsidiaries and headquarters of each bank.

Two departments within the corporate organization of the bank are responsible for the final loan decision: corporate banking and credit risk management. Corporate banking prepares the transaction, negotiates the loan transaction modalities with the corporate client and observes and evaluates the company's business outlook on relevant markets. Risk management performs a corporate evaluation exclusively based on transformed corporate figures. In doing so, risk management functions as "a kind of supervisor" (quoting a department manager) for the corporate banking department. The official and legally coded representation of the loan process then stipulates an organizational work division, the prolongation of the loan preparation and of the decision-making process. This implies that the loan process has two components: on the one side there is the corporate banker (also referred to as the corporate account representative) who has direct contact with the client available, canvasses loans and clients, creates a 'relationship of trust' and brings in a certain amount of self-interest regarding the loan transaction since he or she receives a bonus for each transacted loan. On the other side there is the risk analyst, who is interested solely in the company's written documentation and thus in the construction of the loan.

In this article, practices of calculation are regarded as epistemic practices. Epistemic practices concern the circumstances, events, artifacts, etc., that are taken for granted within the routines of everyday life. At the same time, however, they are themselves routinized practices in their own rights, and their performance is framed by technical devices, procedures, other actors, and negotiations.[1] Consequently the attempt to describe and delineate objects of knowledge is stabilized by technical and other means: they embed these objects in so far as they portray them and make them emerge as such. Needless to say, these technical means are not conceived as neutral objects, but as theory-induced instruments of representation.

In this sense, epistemic practices explore economic objects such as the solvency of a corporation, the development of markets and prices, the dynamics of foreign exchange trading, etc. The banking industry's central object of knowledge—the *economic time* (Kalthoff, 2005, p. 71) of actors and investments—has two implications: firstly, it does not present itself in an unequivocal, immediately recognizable way, but remains vague and even ambivalent, demanding some effort in order to become recognizable. Secondly, the sense of an economic investment is not simply given, but is acquired gradually by the risk management. This article will explore different aspects of economic calculation as it is implemented in risk management. It sketches the general technical framework of risk calculation with regards to global software and banks' data management (§3).

The next section debates and substantiates the assumption that a company's technical devices of calculation constitute the company in the first place. It will also show the constructivity of economic figures that result out of work on the economic category itself (§4). Then I will analyze negotiations among risk analysts in the subsidiaries and headquarters of large international banks concerning those calculated figures. The analysis of the written reports' verbal interpretations aims at shifting the concept of calculation from 'calculating something' to 'calculating with something.' Taking advantage of this relationship between interpretation and representation as well as between speech and writing, the article analyzes practices of communication and cognition in which ideas about and expectations of the future development regarding the bank's economic strategy are involved (§5). Both concepts—'calculating something' and 'calculating with something'—can be traced back to Heidegger's work on the social role and the social function of technology in modern societies. In a first step, Heidegger's position will be closely examined in order to ascertain its relevance for sociology of calculation (§2).

I gathered the empirical material which will be documented in this article through ethnographic fieldwork: about six months of participant observation in two subsidiaries (Warsaw, Sofia) of two international banks:

---

[1] The distinction between objects assumed to be self-evident in routine courses of action and those facing inquiry by means of epistemic practices can also be found in Heidegger (1957b). According to him, instruments are "to hand" (*"zur Hand"*) and are used for specific purposes, to which they themselves unquestionably make reference. But in situations when that object's reference is disrupted, they themselves become questionable and have to be restabilized (cf. Heidegger, 1957b, p. 74).

I observed, among others, social actors in their daily routines of calculation, working on written documents, preparing them for internal negotiations, discussing the results of their calculation on a local and global level. Additionally, I conducted interviews—formal expert interviews as well as "friendly conversation(s)" (Spradley, 1979, p. 80)—in other subsidiaries and in the headquarters of international banks (Prague, Paris, Frankfurt/Main, London, Munich). Whenever the interviewees agreed, the interviews were recorded and transcribed later; when the interviewees disagreed to recording I wrote a protocol afterwards based on my "jottings" (cf. Emerson et al., 1995).[2]

2 Considerations on technology

Within sociology, various approaches examine the role that technology plays and the way it functions. Posthumanist social theories symmetrize the sociological view inasmuch as they assign non-human actors a significant share of the performance of actions. Decentering the subject and the role it plays in sociology is central for these approaches. The invention of technical artifacts to which humans delegate the performance and simplification of functions and options as well as values and routines, etc., dissolves the simple dichotomy between human and non-human actors. Basically, it is about conceiving human agency as a technically framed and performed interaction. *Actor Network Theory* (Latour, 1987) follows a strong, semiotic concept of culture and an analogy between actors and artifacts, which frames the interactive occurrence qua "interobjectivity" (Latour, 1994). In the field of German sociology, Rammert (cf. Rammert, 2008) developed the concept of "technography" within which the concept of "distributed agency" is a key for the empirization and even culturally sociological reformulation of Actor Network Theory. In this approach, connections to social constructivism in STS are drawn, which describe the reciprocal and complex conformities and adjustments between human actors and technical artifacts as well as their integration into social worlds as processes of interpretative stabilization and flexibility (cf. Oudshoorn and Trevor, 2003). In these studies, the autonomy of technical artifacts is confronted with the actors' knowledge of how to implement these for specific purposes without assuming a stable and permanent configuration. Although human actors are framed by technical artifacts, they still determine, invent, and produce them. In recent studies, one may also diagnose a reorientation towards human agency that takes into account the material and technical figuration of human life-worlds (e.g., Pickering, 1995).

---

[2] For a detailed account of ethnographic research and its relation to sociological theory, cf. (Kalthoff, 2003, 2008); on the pitfalls of transcription, cf., e.g., (Hammersley, 2010).

This article pursues a different strategy: in order to explore the logic of (material) practices in risk management departments of international banks, it goes back to Heidegger's work on the philosophy of technology. This involves some difficulty, as his position on technology is somewhat ambivalent: his writings include critical, anti-technological thoughts, belief in human control over technology, and romantic notions about craftsmanship (cf. Dreyfus, 1993, p. 303ff.; Dreyfus, 2002, p. 163f.). In addition, his idiolect virtually invites criticism. The purpose of this article is therefore not to develop a totalizing view of technology and calculation; this could hardly be achieved, neither conceptually nor empirically. Rather, the purpose is to give substance to a constitution theory perspective and thus take up the debate on the relevance of Heidegger's ideas for sociology (cf. Weiss, 2001). This article primarily makes use of Heidegger's later writings and includes interpretations of his work by Dreyfus (1992, 1993, 2002), Dreyfus and Spinosa (1997), Seubold (1986), and other scholars.

2.1 Philosophical Standpoint of Technology

The point of departure in Heidegger's considerations is his dissatisfaction with anthropological and instrumental explanations of technology and technical artifacts. Heidegger states that neither of these positions is wrong; they are, however, inadequate for grasping the essence of technology and for explaining how human beings are involved in technology. According to Heidegger, the essence of technology can only be understood if one considers how something that is absent is brought into existence and thus is being made present. Hence, his primary concern is to understand the essence of technology as a "way of revealing" ("*Weise des Entbergens*"; Heidegger, 1954a, English: p. 12, German: p. 16); "revealing" [*entbergen*] brings forth the invisible or the concealed and therefore into existence. For Heidegger, technology has consequently to be perceived not just as a means, but rather as a "challenging" ("*Herausfordern*"; Heidegger, 1954a, English: p. 14, German: p. 18) of nature, through which, for instance, energy is produced,—exhibiting the revealing essence of technology. In Heidegger's language, artifacts are "set" [*gestellt*] through this challenging, i.e., brought forth and placing the educed artifacts in a different context. For Heidegger, there are three key arguments for understanding the signature of modern technology:

The first argument states that modern technology is like an ordering system ("enframing", Heidegger, 1954a, p. 19; "*Ge-stell*", 1957b, p. 23; Jaeger, 1994, pp. 24–45) that treats every present human or non-human entity as a resource that is used, challenged, and transformed. It is an unquestionable, hardly explainable occurrence that places everything—repeatedly and constantly—in connection with this ordering system (cf. Jaeger, 1994, p. 31). The paradigm for this ordering process and efficiency is energy being "challenged" by a hydroelectric power plant: as a natural object, a river does not produce electrical energy out of itself; the river is set by the power plant, which transforms it to function as a drive for the turbines (cf. Heidegger, 1954a, p. 19f.). The relation between the power plant and the river implies two important strands of argumentation formulated by Heidegger: for one, it shows that every element is set by another element—both human beings as well as non-human entities. This "ordering chain" (Jaeger, 1994, p. 29: "*Kette des Bestellens*"; translation H.K.), which knows no end and is conceived as a circuit, equals an "endless disaggregation, redistribution, and reaggregation *for its own sake*" (Dreyfus and Spinosa, 1997, p. 163, italics in the original). Furthermore, the example also demonstrates that, according to Heidegger, modern technology reduces every object to its material quality and function (e.g., to the generation of hydraulic pressure), making it instantly and utterly available. This feature of modern technology does not exclusively apply to technical hardware, but is equally applicable to information and organizations. In his later writings, the computer-aided processing of information becomes Heidegger's paradigm for technology and modern science (in particular physics; cf. Heidegger, 1957a; Dreyfus and Spinosa, 1997). This notion is associated with the perfection of technology, which shows itself in a "thoroughgoing calculability of objects" (Heidegger, 1957a, p. 121)[3] and therefore becomes "orderable as a system of information" (Heidegger, 1954a, p. 23).[4]

In his second argument, Heidegger suggests a perspective that redefines the subject-object relation: it is not the power plant that has been built into the Rhine River, but the Rhine River that has been built into the power plant. The Rhine River is what it is due to the power plant—and not the other way around. Within the ordering system of enframing, which regards every human being and non-human entity as a resource, modern technology constitutes the world, indicating how objects should be treated and how an effective ordering of resources is organized. Heidegger (1954a, English: p. 17, German: p. 20f.) uses an airplane to illustrate this point: the nature of an airplane cannot be inferred from its material characteristics or its ability to take off, but through the connection into which it has been placed. This is an international system of transport in which human beings serve the purpose of filling the machines that are ready for take-off (cf. Dreyfus and Spinosa, 1997, p. 306). Applied to economic calculation this adds up to the following: the implementation of a calculation aided by technical means stands in a chain of transformations that mobilizes uniformity, reification, and control. With this in mind, economic representation and its technical infrastructure are geared towards producing and circulating economic calcu-

---

[3]"*durchgängige [...] Berechenbarkeit der Gegenstände*" (Heidegger, 1957a, p. 198).
[4]"*als ein System von Informationen bestellbar bleibt*" (Heidegger, 1954a, p. 26).

lations. Comparable to the notion that it is not the wheel that determines rotation, but rotation that determines the wheel (cf. Jaeger, 1994, p. 34), it is assumed here that it is not the risk that causes economic representation and decision, rather it is the calculation and its media that cause the "revealing" of the risk—and thus the market, the business, and the return. This means: modern technology shows us how objects should be treated and how an effective array of resources is organized. And economic calculation does not bring objects that already exist into a visible order; rather, financial objects (such as "cash flow", "EBITDA", "return") do not begin to exist until they have been subject to the process of calculation (cf. Seubold, 1986, p. 87ff.).

Heidegger's third argument extends the concept of calculation beyond dealing with numbers: "To reckon, in the broad, essential sense means: to reckon with something, i.e., to take it into account; to reckon on something, i.e., to set it up as an object of expectation" (Heidegger, 1954b, p. 170).[5] Or he writes: "We take them [the circumstances, H.K.] into account with the calculated intent aimed at specific purposes. We reckon in advance with a specific outcome. [...] This kind of thinking continues to be calculation even if it does not operate with numbers and does not set a mainframe computer going. Calculative thinking computes" (Heidegger, 1959, p. 12; translation H.K.).[6] It is important in this context that Heidegger distinguishes "calculative thinking" [das "rechnende Denken"] from "contemplative thinking" [das "besinnliche Nachdenken"] and "representative thinking" [das "vorstellende Denken"] (Heidegger, 1959, p. 13; translation H.K.). "Calculative thinking" characterizes planning and research (Heidegger, 1959, p. 12) with the objective of being able to precisely know, measure, and define something; "contemplative thinking" is distinctive for the human being as a "meditating being" ["sinnende(s) Wesen"] (Heidegger, 1959, p. 14; italics in the original; translation H.K.) and demands effort and care. "Representative thinking" is an activity that places something else in relation to oneself and structures it according to that representation or imagination; it also bridges the difference between calculative and contemplative thinking (cf. Buckley, 1992, p. 235). In (Heidegger, 1950), Heidegger writes: "'We get the picture' [literally, we are in the picture] concerning something. This means the matter stands before us exactly as it stands with it for us. 'To get into the picture' [literally, to put oneself into the picture] with respect

---

[5]"Rechnen im weiten, wesentlichen Sinne meint: mit etwas rechnen, d.h. etwas in Betracht ziehen, auf etwas rechnen, d.h. in Erwartung stellen" (Heidegger, 1954b, p. 54).

[6]"Wir stellen sie [die Umstände, H.K.] in Rechnung aus der berechneten Absicht auf bestimmte Zwecke. Wir rechnen im voraus auf bestimmte Erfolge. [...] Solches Denken bleibt auch dann ein Rechnen, wenn es nicht mit Zahlen operiert und nicht die Zählmaschine und keine Großrechenanlage in Gang setzt. Das rechnende Denken kalkuliert" (Heidegger, 1959, p. 12). Cf. also (Heidegger, 1957a, p. 168).

to something means to set whatever is, itself, in place before oneself just in the way that it stands with it [...]" (Heidegger, 1950, p. 129).[7] To represent something (in the sense of *vorstellen*) is therefore formulated as "to set out before oneself and to set forth in relation to oneself" (Heidegger, 1950, p. 132).[8] In this concept of picture or image, the activity of producing finds expression via representing (*vorstellen*). To represent means, accordingly, to know something and to have it to hand, to shift relations and thus to structure reality.

As has been shown, Heidegger (1950) emphasizes the empirical relevance of cognitive representations: in this sense, to represent something means to bring it forth by means of this thinking. Wittgenstein (1978) also speaks about the relationship between calculation and assessment and that assessments are configured by calculations, which at the same time are assumed to be stable and unambiguous. "Thus we judge the facts by the aid of the calculation and quite differently from the way in which we should do so, if we did not regard the result of the calculation as something determined once and for all" (Wittgenstein, 1978, p. 325).[9]

2.2 Performation, Performance or Performativity?

But what does this mean for economic sociology in general and the social studies of finance in particular? What do researchers in these areas learn about the practices of economic calculation and computation if one embeds them in a "culture of framing" (Martens, 2001, p. 303)?

Or, in other words, has the theoretical framework of the *Social Studies of Finance* not, for instance, already been mapped out by Actor Network Theory (cf. Callon, 1998)? In more recent research in the area of financial sociology, it was Michel Callon (1998) who made a conceptual suggestion that has been widely received (e.g., MacKenzie, 2003; MacKenzie et al., 2007). Callon essentially develops two arguments: (1) Economic action is embedded in economic theory and its models of economic processes; models of economic theory therefore frame and format economic action. The tendency of economic sociology to observe economic practice and economic theory separately and to treat economic practice as an ontologically independent sphere of the social world is replaced by a symmetrical perspective, which in turn protects (economic) sociology from becoming an ancillary

---

[7]"[W]ir sind über etwas im Bilde. Das will sagen: die Sache selbst steht so, wie es mit ihr für uns steht, vor uns. Sich über etwas ins Bild setzen heißt: das Seiende selbst in dem, wie es mit ihm steht, vor sich stellen und es als so gestelltes ständig vor sich haben" (Heidegger, 1950, p. 82).

[8]"[D]as vor sich hin und zu sich her Stellen" (Heidegger, 1950, p. 85).

[9]"Wir beurteilen also die Fakten mit Hilfe der Rechnung ganz anders, als wir es täten, wenn wir das Resultat der Rechnung nicht als etwas ein für allemal bestimmtes ansähen" (Wittgenstein, 1978, p. 325).

discipline of economic theory that directs its attention toward, for instance, actor's preferences. (2) Callon emphasizes: "*homo oeconomicus* does exist, but is not an ahistorical reality [...] He is the result of a 'process of configuration' and is 'formatted, framed and equipped with prostheses which help him in his calculation...'"(Callon, 1998, pp. 22, 51). Within the context of *Actor Network Theory*, Callon is concerned with the embedding of human actors in a network of non-human means of calculation that have been formatted by economic theory and which allow the actors to perform calculations, to formulate prognoses, and thus evoke actions. Callon's *homo oeconomicus* is therefore one link in a chain of inscriptions (i.e., representations) and socio-technical constellations. Here lies the social location of the "capacity of economics in the performing (or what I call 'performation') of the economy" (Callon, 1998, p. 23). According to Callon, it is not the responsibility of sociology to present a more complex version of *homo oeconomicus*, but to comprehend "his simplicity and poverty" (Callon, 1998, p. 50).

The concept of "performation" (Callon, 1998, p. 23) in particular has recently been questioned (cf. Fine, 2003; Mirowski and Nik-Khah, 2008). This term, which is translated as "performativity" in the Anglo-Saxon discussion can, among others, be traced back to the French branch of research shaping an *économie des conventions*, including authors such as Laurent Thévenot, Robert Salais or Olivier Favereau. This branch of research examines, among other things, how economic actions can be coordinated in such a way that they result in, or their entities are shaped into, a form that is acknowledged as information and thus circulates as a legitimate generalization of particular circumstances (situations, theories, persons, etc.). This is described as "investment in forms" (cf. Thévenot, 1984). Callon's suggestion to speak of a *performation* means formatting an entity through another.[10] To borrow from Latour (1994): economic practice is linked to the laboratories of economic theory by an invisible thread; the formulas developed years ago are maintained and communicated by an industry of programmers, engineers, and managers that determines, channels, and authorizes the framework that again determines the actions performed by economic actors.

In contrast, according to Austin (1992), the cultural theory perspective towards the concept of performativity refers to an entity's realization in and through media (such as language and body) as well as to the performative character of practices that, while they are being performed before an audience, change as they are repeated (cf. Wirth, 2002). The term is

---

[10]Cf. also Heidegger (1957a, p. 124): "Yet while information in-forms, that is, apprises, it at the same time forms, that means, arranges and sets straight". Morgan has studied the 'investment in mathematical forms and models' in more detail (cf., e.g., Morgan, 2001, 2011).

especially important in gender studies (e.g., Butler, 1988), in the production of language (e.g., Krämer, 1996), in the analysis of the performance of self (e.g., Goffman, 1974), as well as in the effect and logic of performative media (cf. Carlson, 1996). Whereas culturally theoretical research is oriented towards microanalytically situating performativity in different contexts of practice or media, Callon's *performation* marks a macrodivision of the world in which the one world (economic theory) penetrates the other world (economic practice). *Actor Network Theory* conceives this penetration as an inscribing translation through which the social, nature, or technology can be transferred into processed writing (e.g., Latour, 1999). This semiotic cultural concept places the things it describes into a different ontological order, while culturally theoretical perspectives of performativity do not perform this reduction. The *performation* theory therefore is distinguished from other discourses on performativity inasmuch as it makes the coding of knowledge in theoretical economic models the point of departure for the analysis of economic practices: what is supposed to be observed is the formatting of the economic world, which corresponds (is supposed to correspond) with the models of economic theory.

With respect to Callon's concept, it should be further noted that it tends to exclude the critical examination of the inconsistencies concerning economic theory (e.g., Cullenberg and Dasgupta, 2001) and thus to curtail the sociological view on the practice of economic theory construction as well as to ignore other social dimensions of economic practices. Nevertheless, the viewpoint taken here is that the framing function of both theoretical economic knowledge (which, for instance, is incorporated in technologies, bank products, and instruments of representation) and governmental regulations (which format the market) are important for sociology of economic action. At the same time, however, it makes sense to rehabilitate the social actors who seemed to disappear in the dynamics of inscription.

In contrast to *Actor Network Theory*, a constitution theory perspective as outlined here—according to Heidegger—is capable of transcending an analysis chiefly stressing the semiotic culture of socio-technical networks.[11] The performance of risk calculation, for example, makes reference to its own performativity, which in turn brings forth economic objects by dint of ordering, categorizing, and computing them. This does not occur independently of human action, but it also does not occur through human action alone. Technical things are means in the hands of human actors, playing an essential part in the way technical things function at the same time (cf. Jaeger, 1994, p. 68; Heidegger, 1954a, p. 16). Looking at it from Heidegger's perspective, both of them are put into context by the process of producing

---

[11] Mitchell (2007) has shown how economics is able to shift the border between informal contexts, non-markets and markets, and thereby create its own objects.

reality. Furthermore, the empirical perspective which is taken up in conjunction with a constitution theory approach allows the actors to act and speak as well as to observe and describe the performativity of economic representations within the scope of risk management as a calculative practice. In this approach, sociological theory construction takes place on the basis of empirical observations whose purpose is not to verify a theoretical model, but to generate and even irritate theory.

3 Technological systems of updating: a framework of risk management

An initial look at the risk management department of a bank shows this department to consist of computers on risk analysts' desks, which are linked up to a computer network. From the perspective of the risk analysts, computers and programs are mere tools that make applications possible. They did not construct the computers, nor did they program the software. They are computer users, i.e., users of software programs which, in fact, have been developed and installed by the responsible group division for risk calculation purposes. While risk analysts are tied to their workstations, their data are mobile and can be moved and transformed by analysts. The activities of risk analysts are individualized, since generally they are responsible for individual companies or branches. Direct professional connections with other risk analysts on a horizontal level are rare; the most contact takes place between the analysts and the corporate account representatives of the respective company or branch.

The computer-based data that the risk analyst can draw on concern companies and branches as well as the economic development of regions and countries. The computer not only makes data available by means of which the analyst can compute economic developments of a company, it also supplies the formats in which this is carried out. Thus cash flow, ratio or projection sheets are specific technologies of economic representation by means of calculation, which in turn require and assert effects of homogenization and simplification.[12] The risk analyst consequently operates on the basis of forms into which his organization has implemented knowledge. This also means that work on the form of calculation is not complete until the moment it is applied; negotiations on the implications of the calculation models or on the implementation of other calculation methods do not take place within risk management but in other bank departments.

Updates of computer programs and the daily backup of information are performed by a backup network into which the banks have integrated their

---

[12]This equally applies for staff whichs works in the areas of FX, bond, or derivative trading or investment banking. Here, computer-based calculation tools for technical analysis are installed onto the computers of the traders who operate with them.

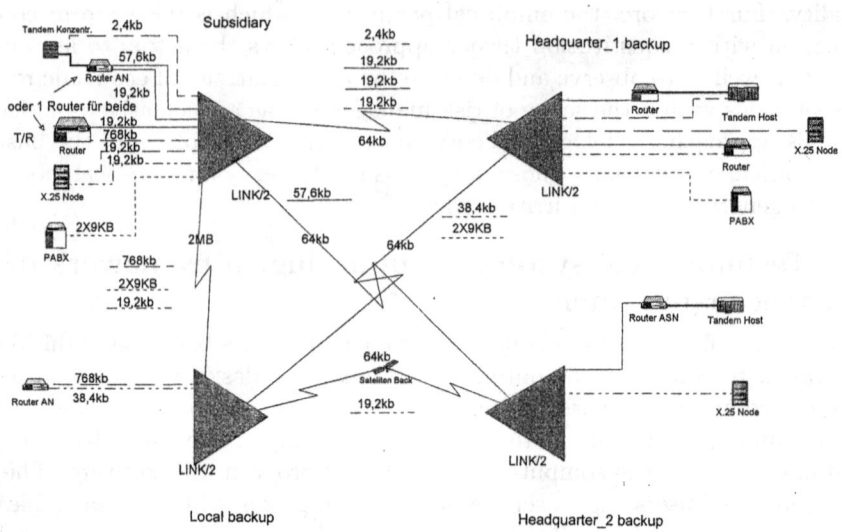

FIGURE 1. Backup Network

subsidiaries. The technical infrastructure of data transfer documented in Figure 1 shows how in the second half of the 1990s, a local internal network (Subsidiary) is linked via routers with the networks of the main data center (Headquarters).

Without being able to fully discuss the technical details here, the following structural traits should be emphasized. The entire backup system of this subsidiary is designed for maximum system stability. This is evidenced by the fact that backups of the entire stock are made at three geographically distributed locations—in a local data center and in two central data centers. Secondly, the way in which the link between the individual networks (Subsidiary, Local backup, etc.) is depicted—using the symbol for lightning ($\lightning$)—indicates that there is no direct connection between the individual networks; rather, there is a connection via an additional medium (telephone network, dedicated line or satellite). It becomes clear that two different kinds of remote data transmissions are used: the organization's own telephone network or dedicated lines between the organizational units, and a satellite connection between "Local backup" and "Headquarter_2 backup". In case of interrupted data transmission pertaining the telephone network or dedicated line, the backup between the local data center and the central data center is secured by a data transmission connection that operates stand-alone, irrespective of the cut off telephone network or the dedicated line. The bandwidth indicated regarding the connection to the data processing centers (Headquarter backups), which from today's point of view seems to be low for the transfer of large volumes of data, refers to mainframe computers in the central data centers that use the network connection more efficiently than conventional computers. Finally, the X.25 node technology is a protocol family developed in the 1970s that enables secure connections via unreliable telephone networks. Even in the 1990s it is plausible to use this technology, as the subsidiary company whose backup network is being represented here operates in Eastern Europe.

It remains to be noted that the updating of computer programs also takes place via these connections: a variety of programs (for acquiring customer data, for carrying out transactions, for calculating and representing economic objects in the various departments, etc.), whose maintenance is also partially delegated to external companies (Bloomberg or Reuters, for instance), are loaded from the central data center via the local data centers onto the individual computers of risk managers and analysts. In the process, regional distinctions are also carried out that take into account the *in situ* requirements and legal conditions. Thus, one program for the logging of customer data provides for 'branch-specific screens' for subsidiaries in New York or Paris.

First, the purpose of this system is to make data globally available in a uniform format and at a wide variety of locations. Secondly, its purpose is not only to make data accessible, but also to enable the data to be processed. This permanent and simultaneous overwriting, backing, overwriting, etc., of data takes place in a simultaneity that makes sure that all of the employees within the local networks who use a specific computer program are up to date and at the same time participating in the generation of the program. Of course—and this is not surprising—there are a variety of computer programs used in the different operative and administrative divisions carrying out the banking business. The following section deals with the use and the performativity of computer-based calculation tools that are implemented in the banking industry for the calculation of creditworthiness.

4 Technology of writing: The constructivity of figures

The calculation and thus the work on the document begins when the company's annual financial statements arrive at the risk management department and the risk analysts manually transfer the bank's balance sheet. Risk analysts refer to this activity as "making a structure". They use it to describe the restructuring and reordering of the company's original balance sheet. In this initial step of transforming figures, individual items are reevaluated, summed up, and rearranged. It often happens, for instance, that the short-

or medium-term liabilities of a company are recalculated or that individual items on the original balance sheet are summed up beneath one item in the structural balance sheet. "Making a structure" therefore means identifying and assigning items as well as performing simple arithmetic operations. In this way, the corporate balance sheet becomes the bank's structural balance sheet; a new company is calculated out of the former company; a self-description becomes a public description. In terms of this study, this activity delineates the reconstitution of the company, since the banking procedure reestablishes its economic framework. With reference to (Foucault, 1975, p. 195ff), it can be pointed out that this activity creates order and methodically places diverging representations into one format: category by category and element by element, the documented wealth of a company is unraveled and rearranged in linear form. This new configuration of economic representation enables recombinations, i.e., different calculations that lead to even further economic representations.

The technical basis regarding the activity of "making a structure" is a calculation program (a so-called data sheet). Here is an example:

Example 1. A risk analyst is sitting in front of his computer and has loaded the input mask into which he wants to enter the new intermediate figures of a company. He clicks on "new customer statement" and enters the date of the annual statement. He then has to opt for an "accounting standard"; he selects "local commercial law" and not the US-GAAP. He begins entering the figures in item 111 ("cash equivalents"). He types the number "662," presses the "return" key, and mutters "zap". The sequence of numbers appears flush-right in the virtual table. The analyst's eyes wander from the original balance sheet to the monitor and from the monitor back to the original balance sheet. This continues category by category, line by line, input field by input field. Several categories are made up of different items in the original balance sheet, thus, e.g., item 321 ("staff expenditures"). The analyst first enters the actual staff costs, to which he then adds the ancillary staff costs and the social welfare expenditure. He controls the accuracy of the input using various items. But the computer also produces warnings. The risk analyst says: "Here's a warning message" and points his finger at a field. He comments further to the ethnographer: "The computer says that the own funds from the previous year plus/minus the total chance in own funds, which results from the overall profit and loss statement, has to result in own funds in the current year".

Example 1 gives an account of a routine activity that can be observed almost on a daily basis. It makes clear that the human actor mediates between a company's written document and the computer, which projects a virtual document onto the monitor. By entering one number after the other, the risk analyst enables the computer to carry out arithmetic operations for which the program is designed. This transfer of the original balance sheet is necessary in order for the computer and the program to interact. At the same time, the company's written document loses its function of supplying the content for the calculation. It becomes superfluous, since it no longer has an influence on the computer's calculation. It is filed away and no longer plays a role in the loan assessment. Thus "making a structure" means transferring the available system of figures of a company into the bank's own scheme and in doing so, generating new documents that represent the central basis for all further calculations, negotiations, and decisions. In this way, restructuring produces the central written banking documents.

Routine activity is interrupted by cases open to question, which has been the case when one risk analyst hesitated when she came across an amount labeled "additional earnings". She assigned the amount to a variable item, i.e., to an item she defined herself, and asked the account representative to clarify this item with the company. The corporate banker told her later that it was an "inter-company loan". In another case the risk analyst was only able to clarify a difference shown on the computer by "trial and error until it worked" and the computer no longer indicated a difference. In concrete terms, one is concerned with establishing the item "fixed assets current year". These "fixed assets" are established through an arithmetic operation: fixed assets (previous year) plus addition to fixed assets minus net disposal of fixed assets minus depreciation results in fixed assets (current year). The problem now does not consist in carrying out this calculation, but lies in knowing which values need to be taken into account for the respective categories (addition to fixed assets, disposal of fixed assets, depreciation).

Even though the calculative steps as well as the values are given, the calculation does not work out. The risk analyst takes a first step and corrects an error. He mistakenly combined depreciation and amortization. He now enters the two numbers separately in the income statement. In a subsequent step he gets hold of a list which contains detailed information about the movement of the fixed assets. He adds those items together that are designated disposal of fixed assets; he then subtracts the depreciation of the disposal of fixed assets from these items and finally gets the amount indicated on the computer (item 612). He proceeds in a similar manner with the additions to fixed assets. The difference identified by the computer came about because the risk analyst initially pursued another philosophy of representation by not categorizing the "down payments for investments" as an "addition". He addressed this issue by saying: "It is often represented very differently. One has to see how it functions by trial and error". This quote articulates a pragmatic attitude towards an applied representation and a tendency to stick to the computer specifications. Furthermore, these scenes show how strongly the computer-based format—in particular the formula fields of the calculation table—intervenes in what is taking place and attracts the risk analysts' attention; they harmonize their ensuing steps according to the format's specifications and calculations, i.e., they add to it, delete numerical values or ignore the "warning". Different participants used the same words to express that they had to make repeated attempts and had to see if they "got it right": "attempting" and "getting it right" clearly show that "making a structure" is not a simple transfer of figures from one calculation scheme to another, but consists of work on fitting those calculation categories since the schemes are based on them. In other words, it shows the constructivity of the figures.

Work on the company's reconstitution can also go so far as to invent corporate figures. In this case, risk analysts constructed a so-called pro forma balance sheet: a balance sheet was subsequently drawn up for a company that had previously belonged to a group of companies and whose balance sheet was not reported independently. According to the risk analyst, the new company's balance sheet was removed from the old company, and they now acted as if the new company had been independent for several years. They were actually operating on the basis of very unstable figures, and they did not really know whether they would be able to get it right.

An indispensable condition concerning the rules of these calculations is the work on the economic category, which ranges from "capital turnover" and "gross turnover surplus" to "FX adjustments". The categories are calculated aided by equations that are often characterized by simple arithmetic operations. The "return on investment" may be cited as an example: The financial equation for "return on investment" is to multiply equity ratio by equity return. Both factors refer to the company's monetary goals, namely "earning money" and "securing the source of earnings" (Baetge, 1998, p. 522); economic discourse draws attention to the earning power and the financial stability of a company, which can be accounted for in this way. From an economic point of view, equity return indicates how efficiently a company has worked; in this respect, it expresses corporate equity. Cash flow, equity capital, and total capital in turn are also fabricated structures, for instance, resulting out of the calculated relation between operating result, standard depreciation (or appreciation), and provision for (or dissolution of) pension accruals (as in the case of cash flow I). The basis of this ramification and interdependency regarding these calculative dimensions form the ratio systems that combine the financial equations with one another and, in the case of the "Dupont Formula", assemble them in the shape of a pyramid to create a "key ratio" (Baetge, 1998). Regarded in computational terms, the equations are created through operations of partitioning, substitution, and extension (cf. Küting and Weber, 2000, p. 27f.). The applied equations produce an effect: they show how economic representations can be treated and combined.

In risk analysis, the calculation of economic figures is first and foremost starting up a calculating machine that has been manufactured and programmed at other locations and that is constantly updated. Employing its calculations, this computer transforms and calculates the empirically existing data material in a way that makes other dimensions of a company available. This work on the document not only minimizes contact with empirical reality, it produces completely new starting points and perspectives. This means: it produces objects in the medium of representation and calculation, and in doing so, presents them to the physical eyes of bank staff. Associated with this is a shift towards the internal plausibility and accuracy of the economic representation.

5 Technology of talk: the structure of economic discourse

As has already been demonstrated, calculative thinking in terms of Heidegger becomes visible within the context of transforming corporate figures and in dealing with requirements specified by computer-based calculation programs. In this section, calculative thinking will be placed into the context of those discussions and examinations that have been conducted with respect to the calculated economic and financial standing of a company and with reference to the transaction decision. As has been explained elsewhere (cf. Kalthoff, 2005, 2011), these verbal discussions take place within and among the participating organizational units—subsidiaries (local level) and bank headquarters (global level). At the local level, the participants are the risk management and corporate banking departments; at the global level, generally, the only participating department is risk management. These discussions have either symptomatic or systematic form. Symptomatic forms include, among others, conversations in the hallway, in passing, or bilateral telephone conversations and e-mail messages. Systematic forms include those kinds of meetings structurally scheduled within a company, such as, e.g., credit committee or telephone conferences. Empirical observations were performed during negotiations in subsidiaries (local level) as well as between the subsidiaries and the bank headquarters (global level) which took place in the form of telephone conversations and telephone conferences.

Each of the locations stands for a different perspective of observation: on the one hand, an acquired (strong) identification with the credit transaction and thus with the company (local side); on the other hand, an observation of this difference, doubt regarding the accurate representation of the figures, and an identification with the bank's specifications (global side). The following remarks highlight the issue of how the actors proceed when they discuss the credit construction and the calculated figures based on the written documents that have been generated. I will elaborate this issue using several examples:

Example 2 (Local risk management 1). The department manager of the risk management department and a staff member (a risk analyst) are sitting in the office of the department manager. The department manager asks the staff member to tell him what speaks in favor of the loan. The risk analyst says that they are not dealing with a steel producer but with a company that processes and coats steel. He adds that over the past several years the company has invested approximately 30 million euros in modernizing the production process. [...] The department manager then asks how things look with the competing companies. The staff member says that steel products from Asia do not constitute competition and that the demand for steel products in Poland is very high. But the department manager interjects that this may be the case today, and points out that compared with Western Europe, steel consumption in Poland is lagging far behind. The risk analyst raises the objection that the company also exports its products to Western Europe. The share may be only ten percent, but the products are also competitive in Western Europe. [...] After both of them spent some time silently leafing through the documents, the staff member says: "Liquidity is a weak point of the company." "Yes, you're right", the department manager agrees.

Example 2 documents a negotiation one often comes across in risk analysis. Bent over a stack of papers, the participants formulate their (critical) inquiries and comments expressing their agreement and their concerns. They articulate what they read, and they listen to what they read but which is articulated by others. In this case, the conversation was prestructured by the fact that both the risk analyst and the department manager—a widely traveled risk specialist for a large bank—basically have a positive attitude towards the loan transaction. This scene documents that the participants proceed according to a question pattern: they inquire into

- the company's economic environment and conditions (for instance, the purchase of steel),
- the market situation (for instance, demand and competition), and
- the company's financial situation based on ratios (for instance, return on equity, liquidity).

The reason why the risk analyst questions these areas is because they are specified in the bank's so-called rating sheet, an evaluation pattern. These items are being formulated during the interaction process and answered on a case-by-case basis. Thus in this interaction process, the account of the assessed company is prestructured by a document that itself is part of the economic evaluation of the company. The items to be assessed throughout the rating sheet are the central theme of the conversation.

But who is speaking here when risk analysts address these items? At a first glance, it is certainly the risk analysts themselves who speak. As they articulate the questions, they are spokespersons of the first order. On the other hand, however, they are also spokespersons of the bank, i.e., of that central department that drafted and tested the risk pattern. In this respect they formulate questions drafted by others concerning this application. They are the spokespersons of risk management.

Example 3 (Local risk management 2). The department manager says: "What kind of a rubbish heap is this? Why are we even doing this? With a 'C,' it's a potential valuation adjustment customer." The risk analyst justifies himself and says: the 'C' is the result of the customer's bad information policy but the transaction itself is safe. And "the company does not have the money at its disposal". The department manager then asks: "And where does the money go?" The risk analyst replies that the money goes into a special account.

Example 3 illustrates the following: a company wants to set up a credit line in order to buy PLN (Polish złoty). The bank's risk is that the company cannot buy two million PLN if the market price for PLN is strong. There are two possibilities for the transaction: first, a spot transaction, which—oriented towards the daily exchange rate—is implemented by the bank's foreign exchange dealers. Secondly, the bank can make use of an FX forward if the amount to be set up is deposited in a special account for which the company has only very limited rights of disposal. In the course of the discussion, a decision is made in favor of this second alternative, and the global markets department is instructed to find another market participant who wants to exchange the amount at an exchange rate of 1.9–2.1 at an appointed time.

The significance of the economic evaluation becomes clear. The bad grade ('C') is offset against a "secure transaction" and thus neutralized. This is accomplished by hedging the transaction in such a way that the loan commitment and the right of disposal are separated. Thus if the economic risk remains manageable and the bank acts as the central player, then a positive decision can result from a bad credit rating.

Example 4 (Between local and global risk management). In the following, we write "Sub" for the subsidiary (local level) and "Hq" for the headquarters (global level).

Sequence 1.

Hq: [...] And so only the new cars are included in the balance sheet.

Sub1: Yes.

Sub2: Yes.

Hq: Because last year they had quite a high capital expenditure, more or less working capital. I assume, to increase their business potential.

Sub2: The capital expenditure is in connection with the cars. Yes, that's right.

Hq: With the cars, okay [...]

Sequence 2.

Hq: [...] We will take an amount of foreign currency receivable from that board?

Sub: Yes.

Hq: Alright. And we will get that portfolio reviewed?

Sub: Sure.

Hq: Okay. Then these payments will be made into accounts in Sofia or in the branches?

Sub: Yes, in Sofia, because we have all the accounts technically here. Everything is done over here.

Hq: Okay. And if there are receivables which are not being paid we would have the right to replace these with other receivables of our choice?

Sub: Yes of course.

Hq: Okay [...].

Sequence 3.

Sub: [...] Something else is said about the leverage ratios on page 24. Good, on the whole that's natural, total liabilities versus equity. We have to make sure that this company stands at the lower end with respect to ratio. I mean this in a positive way. At the lower end if we compare it with the entire industry. And we have a projection that sees an increasing reduction of this leverage ratio. And then there will be a capital increase of 290 million złoty. Then there's somewhat more space again.

Hq: Yes, but that's the overall leverage situation. You know, we have a country lore that says ten percent of the equity capital. Those are our boundaries. And that is clearly above that.

Sub: Wait a minute [he is using a pocket calculator and goes through some documents]. Let me have a look. We're at eighteen percent. And that's too high for you?

Hq: Yes. And it's primarily this loan. We don't want to get rid of any plain vanilla loan products [...]

The three sequences in Example 4 demonstrate another scheme. The risk managers from bank headquarters inquire into details about the calculation, management, the market situation, and the transaction and, as the third sequence documents, also cast doubt on the calculation that has been performed. What is important here is that to them, the economic account presented by the local side in the written documents is not evident. They approach the representation with an element of doubt regarding the accurate portrayal of the company's situation as outlined by the local side. By reviewing the economic account, they check the soundness of the argumentation, the accuracy of the calculation, as well as the hedging of the transaction. This can also lead to the deconstruction of the local side's calculation model. What also becomes obvious in the third sequence is the discrepancy between local market and global strategy: the spokeswoman for the global side reminds the local risk analysts of the general business strategy not to set one's stakes on simple investments and bank transactions but on complex ones. The sequences also highlight the hierarchical structure of the discourse: in the first two sequences, a local department manager and a risk analyst justify themselves to a higher-level manager from the bank's headquarters (global risk management). Similar to a courtroom situation, their expressions of confirmation are brief.

The scenes documented throw light on the systematic elements concerning processes of consensus that are carried out in a large international bank on the basis of written documents: (1) the prestructuring of the discourse through an assessment instrument which directs the actors' attention; (2) the non-economic legitimization of economic decisions, that is, the discursive use of non-economic figures of argumentation; (3) the deconstruction of economic calculations and thus the argumentative preparation of grounds that can lead to a rejection of the loan application; and (4) finally, doubt with regard to the accuracy of the portrayal, which leads to detailed inquiries. These four elements recur in the verbal negotiations in various constellations and variations. Furthermore, they document which form the social phenomenon of "calculating with something" assumes in this area.

6 Conclusion

This article dealt with the question of how sociology of calculation can be worked out using the example of economic calculation and thus empirically analyzing it. We have argued that the impetus for sociology—but also the particular challenge—consists in taking up a constitution theory perspective in which it is assumed that (economic) entities can be produced by ways of calculative methods and processes. Conceiving the implementation of economic calculations as the (technically based) constitution of entities and as a reciprocal relationship between the world of figures and the imagination has consequences for sociological research: first, the mathematical realism characteristic of, e.g., work in the area of sociology of knowledge (e.g., Mannheim, 1929; Bloor, 1973) has to be rejected. Mathematical realism assumes the existence of entities that are only reordered and portrayed by calculation processes. In contrast, the constitution theory perspective suggested here argues that the written processes of mathematical operations produce the objects of the economy; it is the processes of calculation that cause these entities (e.g., cash flow, EBITDA, risk) to exist in the first place. For the sociology of economics and finance, the development, implementation, and use of these "operative modes of writing" (Krämer, 1997) marks an empirical research program.

Secondly, the theory of visibility as formulated by accounting research (e.g., Hopwood and Miller, 1994) needs to be complemented. It is argued in these studies that hidden elements of the economy can be made visible using a scopic technology of calculation, and that exercising control becomes effective through this process of being made visible (cf. Foucault's model of panoptism; Foucault, 1975).[13] What these studies neglect is the other side of visibility, viz., a vision that is equipped with power and knowledge and that is directed toward (self-)knowledge. It is important for sociology of calculation to tackle the elaboration of a theory concerning this kind of economic "worry about oneself" (Foucault, 2004) as an independent mechanism of calculative practice.

Thirdly, the concept of practice requires clarification and conceptualization. As a rule, in the areas of economic sociology or accounting studies it is accompanied by a concept of action that implies individual, goal-oriented actors with intentions. This kind of concept of action only allows a limited view of issues concerning the practice of calculation, the constructivity of sets of figures, the function of technical artifacts, or the role of the human body. Thus an advance is being made for a culturally sociological concept that does not individualize action, but locates it in the performance of social practice, which in turn exhibits physical and technical, representational

---

[13]The term "scopic technology" loosely follows Foucault (1975); for the notion of "scopic system", cf. Knorr (2006).

and reflexive dimensions. There are culturally sociological concepts available that might serve as the basis for follow-up research.

The research perspective suggested does not imply a (neo-)Kantian turn in the sociology of calculation or a revitalization of the old debate over realism versus relativism with reference to the reality of economic entities or facts. Rather, it perceives work on the written document and the calculation as a practice of representing in its own right. In this way it follows observable practices and empirical relations; it implies an awareness of the calculative framework and thus the question of what is included in or excluded from the calculation through work on the category.

Bibliography

Austin, J. L. (1992). *How to do things with words*. Oxford University Press, New York.

Baetge, J. (1998). *Bilanzanalyse*. IDW-Verlag, Düsseldorf.

Beunza, D. and Stark, D. (2004). Tools of the trade: The socio-technology of arbitrage in a Wall Street trading room. *Industrial and Corporate Change*, 13(2):369–401.

Bloor, D. (1973). Wittgenstein and Mannheim on the sociology of mathematics. *Studies in History and Philosophy of Science*, 4:173–191.

Buckley, R. P. (1992). *Husserl, Heidegger and the crisis of philosophical responsibility*. Kluwer, Dordrecht.

Butler, J. (1988). Performative acts and gender constitution. An essay in phenomenology and feminist theory. *Theatre Journal*, 40(4):519–531.

Callon, M. (1998). Introduction: The embeddedness of economic markets in economics. In Callon, M., editor, *The laws of the market*, pages 1–57, Oxford. Blackwell.

Carlson, M. (1996). *Performance: A critical introduction*. Routledge, London and New York.

Cullenberg, S. and Dasgupta, I. (2001). From myth to metaphor. A semiological analysis of the Cambridge capital controversy. In Cullenberg, S., Amariglio, J., and Ruccio, D. F., editors, *Postmodernism, economics and knowledge*, pages 337–353, London and New York. Routledge.

Dreyfus, H. (1992). Heidegger's history of the being of equipment. In Dreyfus, H. L. and Hall, H., editors, *Heidegger: A Critical Reader*, pages 173–185, Oxford. Blackwell.

Dreyfus, H. L. (1993). Heidegger on the connection between nihilism, art, technology, and politics. In Guignon, C. B., editor, *The Cambridge Companion to Heidegger*, pages 289–316, Cambridge. Cambridge University Press.

Dreyfus, H. L. (2002). Heidegger on gaining a free relation to technology. In Dreyfus, H. and Wrathall, M., editors, *Heidegger reexamined. Volume 3. Art, poetry, and technology*, pages 163–174, New York and London. Routledge.

Dreyfus, H. L. and Spinosa, C. (1997). Highway bridges and feasts: Heidegger and Borgmann on how to affirm technology. *Man and World*, 30(2):150–177.

Emerson, R. M., Fretz, R. I., and Shaw, L. L. (1995). *Writing ethnographic fieldnotes*. University of Chicago Press, Chicago, IL.

Fine, B. (2003). Callonistics. A disentanglement. *Economy and Society*, 43(3):478–484.

Foucault, M. (1975). *Surveiller et punir. Naissance de la prison*. Gallimard, Paris. Page numbers refer to the English translation (Foucault, 1995).

Foucault, M. (1995). *Discipline & punish: The birth of the prison*. Vintage Books, New York.

Foucault, M. (2004). *The hermeneutics of the subject. Lectures at the Collège de France 1981–1982*. Palgrave Macmillan, New York.

Goffman, E. (1974). *Frame analysis: An essay on the organization of experience*. Harper & Row, New York.

Hammersley, M. (2010). Reproducing or constructing? Some questions about transcription in social research. *Qualitative Research*, 10:553–569.

Hegel, G. W. F. (1807). *Phänomenologie des Geistes*. J.A. Goebhardt.

Heidegger, M. (1950). Die Zeit des Weltbildes. In Heidegger, M., editor, *Holzwege*, pages 69–104, Frankfurt a.M. Klostermann. Page numbers of English quotations refer to Heidegger (1977a).

Heidegger, M. (1954a). Die Frage nach der Technik. In Graf Podewils, C., editor, *Die Künste im technischen Zeitalter*, volume 3 of *Gestalt und Gedanke*, pages 70–108, München. Oldenbourg. Page numbers refer to Heidegger (2000, pp. 9–40), page numbers of English quotations refer to Heidegger (1977b).

Heidegger, M. (1954b). Wissenschaft und Besinnung. In Heidegger, M., editor, *Vorträge und Aufsätze*, pages 45–70, Pfullingen. Neske. Page numbers refer to Heidegger (2000, pp. 41–66); page numbers of English quotations refer to Heidegger (1977c).

Heidegger, M. (1957a). *Der Satz vom Grund*. Neske, Pfullingen. Page numbers refer to the 8th edition, 1997; page numbers of English quotations refer to Heidegger (1991).

Heidegger, M. (1957b). *Identität und Differenz*. Neske, Pfullingen. Page numbers of English quotations refer to Heidegger (2002).

Heidegger, M. (1959). *Gelassenheit*. Neske, Stuttgart. Page numbers refer to the 12th edition, 2000.

Heidegger, M. (1977a). The age of the world picture. In Heidegger, M., editor, *The question concerning technology and other essays*, pages 115–154, New York. Harper & Row.

Heidegger, M. (1977b). The question concerning technology. In Heidegger, M., editor, *The question concerning technology and other essays*, pages 3–35, New York. Harper & Row.

Heidegger, M. (1977c). Science and reflection. In Heidegger, M., editor, *The question concerning technology and other essays*, pages 155–182, New York. Harper & Row.

Heidegger, M. (1991). *The principle of reason*. Indiana University Press, Bloomington, IN.

Heidegger, M. (2000). *Vorträge und Aufsätze*. Neske, Stuttgart, 9th edition.

Heidegger, M. (2002). *Identity and Difference*. University of Chicago Press, Chicago, IL.

Hopwood, A. G. and Miller, P., editors (1994). *Accounting as a social and institutional practice*. Cambridge University Press, Cambridge.

Jaeger, P., editor (1994). *Martin Heidegger. Gesamtausgabe. III. Abteilung: Unveröffentlichte Abhandlungen. Band 79: Bremer und Freiburger Vorträge*. Klostermann, Frankfurt a.M.

Kalthoff, H. (2003). Beobachtende Differenz. Instrumente der ethnografisch-soziologischen Forschung. *Zeitschrift für Soziologie*, 32:70–90.

Kalthoff, H. (2005). Practices of calculation. Economic representation and risk management. *Theory, Culture & Society*, 22(2):69–97.

Kalthoff, H. (2008). Zur Dialektik von qualitativer Forschung und soziologischer Theoriebildung. In Kalthoff, H., Hirschauer, S., and Lindemann, G., editors, *Theoretische Empirie. Die Relevanz qualitativer Forschung*, pages 8–38, Frankfurt a.M. Suhrkamp.

Kalthoff, H. (2011). Un/doing calculation. On knowledge practices of risk management. *Distinktion. Scandinavion Journal of Social Theory*, 12(1):3–21.

Kalthoff, H., Rottenburg, R., and Wagener, H.-J., editors (2000). *Facts and figures. Economic representations and practices*. Metropolis, Marburg.

Knorr, K. (2006). The market. *Theory, Culture & Society*, 23(2/3):151–156.

Knorr, K. and Brüger, U. (2002). Global microstructures: The virtual societies of financial markets. *American Journal of Sociology*, 107:905–950.

Krämer, S. (1996). Sprache und Schrift oder: Ist Schrift verschriftete Sprache? *Zeitschrift für Sprachwissenschaft*, 15:92–112.

Krämer, S. (1997). Kalküle als Repräsentation. Zur Genese des operativen Symbolismus in der Neuzeit. In Rheinberger, H.-J., Hagner, M., and Wahrig-Schmidt, B., editors, *Räume des Wissens*, pages 111–122, Berlin. Akademie Verlag.

Küting, K. and Weber, C.-P. (2000). *Die Bilanzanalyse: Lehrbuch zur Beurteilung von Einzel- und Konzernabschlüssen*. Schöffer-Poeschel Verlag, Stuttgart.

Latour, B. (1987). *Science in action: How to follow scientists and engineers through society*. Open University Press, Milton Keynes.

Latour, B. (1994). Une sociologie sans objet? Remarques sur l'interobjectivité. *Sociologie du Travail*, 36:587–607.

Latour, B. (1999). *Pandora's hope. Essays on the reality of science studies*. Harvard University Press, Cambridge, MA.

MacKenzie, D. (2003). An equation and its worlds: Bricolage, exemplars, disunity and performativity in financial economics. *Social Studies of Science*, 33:831–868.

MacKenzie, D. (2006). *An engine, not a camera. How financial models shape markets*. MIT Press, Harvard.

MacKenzie, D., Muniesa, F., and Siu, L., editors (2007). *Do economists make markets? On the performativity of economics*. Princeton University Press, Princeton, NJ.

Mannheim, K. (1929). *Ideologie und Utopie*. F. Cohen, Bonn.

Martens, W. (2001). Ist das Gestell funktional differenziert? Heideggers Technikverständnis und Luhmanns Theorie funktional differenzierter Gesellschaften. In Weiss, J., editor, *Die Jemeinigkeit des Mitseins. Die Daseinsanalytik Martin Heideggers und die Kritik der soziologischen Vernunft*, pages 295–326, Konstanz. UVK.

Mirowski, P. and Nik-Khah, E. (2008). Command performance: Exploring what STS thinks it takes to build a market. In Trevor, P. and Swedberg, R., editors, *Living in a material world: Economic sociology meets science and technology studies*, pages 89–128, Cambridge, MA. MIT Press.

Mitchell, T. (2007). The properties of markets. In MacKenzie, D., Muniesa, F., and Siu, L., editors, *Do economists make markets? On the performativity of economics*, pages 244–275, Princeton, NJ. Princeton University Press.

Morgan, M. S. (2001). Making measuring instruments. In Klein, J. L. and Morgan, M. S., editors, *The age of economic measurement*, pages 235–251, London. Duke University Press.

Morgan, M. S. (2011). *The world in the model*. Cambridge University Press, Cambridge.

Oudshoorn, N. E. and Trevor, J. P., editors (2003). *How users matter. The co-construction of users and technology*. MIT Press, Cambridge, MA.

Pickering, A. (1995). *The mangle of practice. Time, agency, and science*. University of Chicago Press, Chicago, IL.

Rammert, W. (2008). Technographie trifft Theorie. Forschungsperspektiven einer Soziologie der Technik. In Kalthoff, H., Hirschauer, S., and Lindemann, G., editors, *Theoretische Empirie. Die Relevanz qualitativer Forschung*, pages 341–367, Frankfurt a.M. Suhrkamp.

Seubold, G. (1986). *Heideggers Analyse der neuzeitlichen Technik*. Alber, Freiburg, München.

Spradley, J. P. (1979). *The ethnographic interview.* Holt, Rinehart and Winston, New York.

Thévenot, L. (1984). Rules and implements: Investment in forms. *Social Science Information*, 23:1–45.

Weiss, J., editor (2001). *Die Jemeinigkeit des Mitseins. Die Daseinsanalytik Martin Heideggers und die Kritik der soziologischen Vernunft.* UVK, Konstanz.

Wirth, U. (2002). Der Performanzbegriff im Spannungsfeld von Illokution, Iteration und Indexikalität. In Wirth, U., editor, *Performanz. Zwischen Sprachphilosophie und Kulturwissenschaften*, pages 9–60, Frankfurt a.M. Suhrkamp.

Wittgenstein, L. (1978). *Remarks on the foundations of mathematics.* Basil Blackwell, Oxford, revised edition. Page numbers refer to the 1996 edition; page numbers of German quotations refer to the German edition (Wittgenstein, 1984).

Wittgenstein, L. (1984). *Bemerkungen über die Grundlagen der Mathematik*, volume 506 of *suhrkamp taschenbuch wissenschaft*. Suhrkamp, Frankfurt a.M.

François, K., Löwe, B., Müller, T., Van Kerkhove, B., editors,
Foundations of the Formal Sciences VII
Bringing together Philosophy and Sociology of Science

Demystification of early Latour

JOUNI-MATTI KUUKKANEN*

Instituut voor Wijsbegeerte, Universiteit Leiden, Postbus 9515, 2300 RA Leiden, The Netherlands
E-mail: j.kuukkanen@phil.leidenuniv.nl

There are two conclusions that one can safely draw from the debates on the relationship between sociology of science and philosophy during recent decades: The sociologists of science have been typically perceived as advocating social constructivism, and philosophers have generally attacked this position as indefensible or even bizarre. My intention in this paper is to muddy the waters and show that the issue is far from this simple. I concentrate on Bruno Latour, who is usually taken as one of the most extreme scholars in an already radical constructivist camp.[1] He has become famous for his wide-ranging constructivist theses, which include the claim that facts and reality are scientists' constructions.

This paper is an attempt to offer a common-sense reading of Latour with the conceptual machinery of analytic philosophy. More specifically, my paper intends to show that despite an appearance to the contrary Latour ought not to be taken as a metaphysical social constructivist in the sense that the philosophers of science use this label. If this perspective strikes one as being insensitive to the kind of scholarship Latour represents, my defence is that this is an attempt to translate Latour's discourse into a language that analytic philosophers also understand, in which Latour's statements are interpreted as authentically as possible. One hopes that this endeavour might help to enhance communication across various disciplines in science studies, from sociology to philosophy of science and vice versa.

---

*I would like to express my sincere gratitude to James McAllister for his constructive criticism. I also thank another member of our research group *Philosophical Foundations of the Historiography of Science*, Bart Karstens, for his helpful feedback on the final draft. Finally, I thank anonymous referees for their useful comments. Needless to say, no one else but me should be held responsible for any remaining shortcomings in the paper.

[1] A representative selection of various sociological approaches on science can be found in Pickering's (1992) edited book *Science as Practice and Culture*.

Received by the editors: 22 January 2009; 9 October 2009.
Accepted for publication: 28 October 2009.

My focus is primarily on Latour's earlier work. It offers a fruitful object of study, having generated many famous wide-ranging constructivist theses. The key text is Latour and Woolgar's groundbreaking *Laboratory Life. The Social Construction of Scientific Facts*, first published in 1979. Another central piece is Latour's first general presentation of his network model, *Science in Action*, which appeared in 1987. *Pasteurization of France* that was published in French a couple of years earlier and was translated into English in 1984 provides a good illustration of how Latour applies his network model to one well-know episode in the history of science, the scope of which also is historically much wider than that of *Laboratory Life*. Finally, I have used Latour's *Pandora's Hope* from 1999 as a further exemplification of the themes in his scholarship and as Latour's self-commentary on his earlier work. Latour has naturally published other books and articles afterwards, but they appear in this essay only in passing for two reasons. My intention is not to conduct a comprehensive exposition on Latour's whole career, which would not only be a challenging task to accomplish, but beyond the scope of an article length publication. Second, the early works form a sufficiently self-contained and an interesting object of research as such. I should not attempt to evaluate whether the analysis offered is applicable to all Latour's later works, but it certainly is a possibility that cannot be discounted.

Latour draws strict limits around the conceptual resources that may be used in studying science; therefore, it is interesting to ask whether within these limits we are able to give a satisfactory account or 'theory' of scientific activity. I will highlight both a more deflationary, and a more positive side of his philosophy, because my intention is to divert philosophers attention to the questions where one expects to find proper disagreements between Latour and most philosophers of science. This is to say, philosophers' engagement with (at least the Latour kind of) sociologists should focus on the question of how well the latter manage to explain the advancements and failures of science. The question is important, as it may help us to understand the limits (as well as benefits) of the kinds of explanation offered by sociologists. This examination also is well-justified in the current intellectual climate, because while Latour's science studies is often regarded as absurd by philosophers of science, it has fallen on fertile ground among many contemporary historians of science (cf. Golinski, 1988).

In this paper, I will first briefly consider different kinds of social constructivism and consider their relevance to our examination. Latour's various constructivist claims seem and are usually taken to fall in the category of metaphysical constructivism. However, I will show that he is not a metaphysical constructivist and that his constructivist statements are somewhat trivial claims. His theses become comprehensible also from a philosopher's point of view once one remembers that Latour studies science as an anthropologist. This demystification of Latour via his anthropological perspective on science is then continued by taking a look at other more specific constructivist statements that can be found in his writings, such as the claim that Pasteur constructed microbes. After that I examine Latour's 'positive' explanatory model trying to find the most fundamental explanatory principles in his theory. It turns out that the restrictions put forward by Latour on the kinds of notion permissible in the explanation of science stem from what philosophers would call epistemological anti-realism. For example, he denies not that there is reality but that one can have independent access to it. This examination ends with a sceptical remark on whether Latour is able to explain satisfactorily success and failure in science.

1 Social constructivism

Sociologists of science have been accused of various kinds of social constructivism by philosophers of science. It is therefore important to pay attention to how we understand 'social constructivism' and narrow the scope of focus accordingly. Paul Boghossian (2001) has distinguished two different theses of social constructivism that often cause controversy. The first is the metaphysical claim according to which something is real but of our own creation. The second claim is an epistemological one which says that the reason for having a particular belief boils down to the role this belief plays in our social lives rather than to the evidence in its favour. André Kukla has approached social constructivism slightly differently in his *Social Constructivism and the Philosophy of Science* (2000). He distinguishes three types of social constructivism: metaphysical, epistemological and semantic. The metaphysical thesis says that the objects of science are invented or made, rather than discovered. The epistemological thesis commits one to a view that there is no absolute warrant for any belief; any rational warrant is relative to a culture, individual or paradigm. The semantic thesis in turn means that sentences do not have determined empirical content because they are not appropriately connected with the world, and so any verbal outcome is subject to negotiation. Kukla emphasises that all these types of constructivism are independent of each other[2]. To take one more example, Hacking defines social constructivism more broadly, by three "sticking points": contingency, nominalism and explanations of stability. Seen from the social constructivist point of view, the position boils roughly down to three statements: Objects in science need not have existed at all, the world we describe does not have any pre-given structure and the explanations for the stability of scientific beliefs involve external elements to the content of science Hacking (2001, Chapter 3).

---

[2]Kukla (2000, pp. 5–6); cf. Chapters 1 and 2 for further references on social constructivism.

The most radical form of social constructivism is arguably the one which says that the objects themselves are created or invented in science, and thus not discovered by scientists, i.e., metaphysical social constructivism. Yet, the radicality of this thesis depends on its domain of application. It is not striking to claim that beliefs or science are constructed. Both science and scientific beliefs are human constructions in a trivial sense. It is not either revolutionary to say that *some* objects of scientific research are constructed, because there clearly are objects which are not found independently in nature. The periodic table contains over 20 elements that are not found in nature and must thus be synthesized by humans.[3] Some of Latour's claims about construction seem, even at the first sight, to fall under this trivial construction category. Bioassay or any phenomena whose production depends on it are clearly constructed by scientists in their laboratories (cf. Latour and Woolgar, 1979, p. 64).

One also needs to be sensitive to the fact that Latour and Woolgar question the applicability of the term 'social construction'. In the postscript of the second edition of *Laboratory Life*, they explain their reason to drop the term 'social' from the title of the book, the consequence of which is that *Laboratory Life. The Social Construction of Scientific Facts* became *Laboratory Life. The Construction of Scientific Facts*. Latour and Woolgar write that the use of the term was ironic in the first place and that it denotes too broad a category, making the term practically meaningless. The point is that "*all* interactions are social" (Latour and Woolgar, 1979, p. 281; original emphasis) and so it is better to just talk about 'construction of scientific facts'. In light of some later writings in which Latour tries to overcome dichotomous explanations that rely on a one-dimensional axis whose poles are social and nature, it is surprising to deem all interactions social (e.g., Callon and Latour, 1992; Latour, 2005, p. 254). By doing this Latour seems to legitimate the practice of labelling him as 'social constructivist'. However, this would be an incorrect judgment, because on various occasions Latour stresses that non-human entities have a major role to play in the process of construction and that they can be taken as 'social actors' as well (e.g., Latour, 2005, pp. 92 & 106–107). Indeed, in *Science in Action*, Latour introduced the notion 'actant' that does not imply any *a priori* distinction between human and non-human factors.

Further, in a relatively recent publication, *Reassembling the Social*, Latour explicitly recognises the ambivalent message that the talk of 'social construction' and 'construction' more generally conveys, i.e., that most commentators of his works took it that "*either* something was real and not con-

---

[3] An 'objectivist' would probably point out that these elements are natural kinds, which exist independently of humans. This is however not important here, because the point is that the talk of construction in science makes sense if it is appropriately qualified.

structed, or it was constructed and artificial, contrived and invented, made up and false" (Latour, 2005, p. 90; original emphasis). Latour's point is not to cast doubt on the reality of objects, but to offer an account of the contingent process that underlies the emergence of entities in the ontology of science: "When we [actor network scholars] say that a fact is constructed, we simply mean that we account for the solid objective reality, by mobilizing various entities whose assemblage could fail" (Latour, 2005, p. 91; original emphasis).

It is therefore important to realise that being constructed is not the same as being unreal or less real, as one may be tempted to think. A telephone is not imaginary or fiction although constructed. The reality of artificially derived elements would become very clear to anybody who come to contact with them, as they tend to be radioactive. It seems that the actual issue behind metaphysical constructivism is whether an object is created by humans (perhaps in cooperation with non-human entities) or whether it is part of the human-independent ready-structured world. Common sense seems to dictate that a telephone needs human construction in order to come into existence, but a stone does not. But both are real. For this reason, the question that we have to pose to Latour is whether he maintains that the objects of science are made by humans or whether they are out there independent of human beings irrespective of the question of their reality (cf. Hacking, 2001, Chapter 5).

Latour claims that there was no microbe to be discovered by Pasteur and early microbiologists: it was just a temporary construction which eventually evaporated (Latour, 1984, p. 108). He also makes clear in *Laboratory Life* that, with the help of some material instrumentation, the hormone TRF was made by scientists in the laboratory (e.g., pp. 64, 125–126, & 176–177). Even better, Latour claims that facts in general are constructed and that the representation of nature is the consequence of the settlement of a controversy, not the other way round (e.g., his third rule of method, see Latour, 1987, p. 99).

Now this is something else. The view above has been dismissed equally by natural scientists and philosophers of science. Practically all the commentators in these fields are puzzled how anyone can defend such a position. Hacking writes that Latour is taken almost universally as a "constructionist", and probably because of this position, many regard him as the public enemy number one (Hacking, 2001, pp. 64–65; cf. Kukla, 2000, p. 9; Newton, 2000, p. 200). Boghossian writes suggestively that it is not easy to make sense of the idea that facts about elementary particles or dinosaurs are a consequence of scientific theorizing (Boghossian, 2001). Indeed, it is not, but I intend to show next that it is nevertheless possible.

2   Anthropological standpoint

In order to comprehend Latour's constructivism we have first to understand his approach to science. Fortunately, he is quite explicit about it. In *Laboratory Life*, Latour and Woolgar define what is new and specific in their study of science. Their concern is with daily activities of working scientists, or as they call it, 'the soft underbelly of science' (Latour and Woolgar, 1979, pp. 27 & 20). The investigators focus on "observations of actual laboratory practice" (Latour and Woolgar, 1979, p. 153) or draw attention to "the process by which scientists make sense of their observations" (Latour and Woolgar, 1979, p. 32). Further, the authors compare their orientation to that of an anthropologist and describe their point of view as ethnographic. More specifically, they utilise the idea of "anthropological strangeness", according to which the lack of prior knowledge does not prevent one from gaining understanding of science (Latour and Woolgar, 1979, p. 29; Latour, 1984, pp. 125–126).

Latour's approach reminds us of Quine's radical translator who studies a native tribe without any prior knowledge of its habits or language, except that the authors of *Laboratory Life*, of course, have some prior understanding of the functioning of science. They have to therefore pretend not to understand and concentrate only on what they see with their own eyes in the laboratory. All in all, the task is to understand science and scientist's activities by direct observation and without any presumptions as to what constitutes and explains scientific activity.

This approach to science imposes serious methodological limitations for permissible action by the investigators. By paying attention to 'routinely occurring minutiae' (Latour and Woolgar, 1979, p. 27) one won't come across 'truth' or 'reality' in action in science. Since our anthropologist enters the laboratory without preconditions of what constitutes knowledge (Latour, 1987, p. 13) and without the pre-empirical presumption that scientific knowledge is qualitatively different from common sense, the authors refrain from using any epistemological concepts in their explanations (Latour and Woolgar, 1979, p. 153). One can thus see that the aim of the anthropological approach is to follow the action of scientists as close to the surface level as possible, at the immediate observable social level, without using any context-transcending notions in explanations and without making any assumptions prior to investigation. Further, any context-transcending explanation or employment of context-transcending concepts in one's explanations need to be justified empirically. Latour's method might well be called extreme empiricist and descriptivist.[4]

---

[4] Cf. Collins and Yearley, 1992.

3 Social construction deconstructed

The role and status of facts make a rewarding topic of analysis in Latour, because Latour is explicit and offers us a clear and direct definition of 'fact'. "A fact is nothing but a statement with no modality—M—and no trace of authorship" (Latour and Woolgar, 1979, p. 82). Latour describes five different stages of fact-making. At the lowest level, 'fact' has actually the status of an 'artefact'. Facts on this level are typically speculative statements or comprise conjectures by individual scientists, and appear at the end of papers or in private discussions. Type 2 statements contain modalities that draw attention to the generality of available evidence in favour of or against the statements. Type 3 statements include specific references to the evidence for the statements. Modality has been removed from statements of type 4, but they contain references. Finally, statements of type 5 are 'facts', because they are devoid of modalities and qualifications; they are straightforward statements of the state of affairs. What scientists try to do, according to Latour, is to get their colleagues to drop all the modalities from statements that they originated, which is to say that they try to make a statement more of a fact, and consequently, less of an artefact (Latour and Woolgar, 1979, pp. 91–82; cf. Latour, 1987, p. 42–44).

It is worth pausing at this point and spelling out clearly what Latour is getting at. He is talking about *statements* (and sometimes about sentences). It is hardly radical to insist that any statement was uttered somewhere at some specific point of time. So when Latour says that the fate of facts is in the hands of later users (e.g., Latour, 1987, p. 38) or that the construction of facts is a collective process (e.g., Latour, 1987, p. 29), this ought not to be surprising. Removing modalities from a statement that was a highly speculative conjecture and its incorporation into an encyclopaedia as fact knowledge certainly is dependent on numerous other scientists and experts in the field. Further, when he says that statements rarely achieve the status of 'fact', he must be absolutely right (Latour, 1987, p. 42). Most statements made won't ever be taken for granted or accepted as such. But if that happens, then 'fact' has been constructed in and by a community.

Latour's statements about 'reality' can be interpreted in a similar manner. He writes,

> Laboratories are now powerful enough to define reality... [R]eality as the Latin word *res* indicates, is what resists. What does resist? Trials of strength. If, in a given situation, no dissenter is able to modify the shape of a new object, then that's it, it is reality, at least for as long as the trials of strength are not modified... The minute the contest stops, the minute I write the word 'true', a new, formidable ally suddenly appears in the winner's camp, an ally invisible until then, but behaving now as if it had been there all along: Nature. (Latour, 1987, pp. 93–94)

How could laboratories define 'reality' or 'nature'? The answer: in the case where we are talking about 'reality' or 'nature' on the level of a scientist's action. Latour really studies *science in action* and one of the results of this process is 'reality' and 'nature'. Philosophers would probably be inclined to call them conceptions of what is real or nature, which is fine as long as one remembers that such notions are not allowed in Latour's theoretical model. Morphine, endorphin, the physiograph (which enables a graphical presentation of gut pulsation) and so on become 'real' or (part of) 'nature' when they are 'blackboxed', i.e., they become so well defined and tested that an individual scientist is no longer able to question them. 'Nature' or 'reality' could be understood as a collection of all taken-for-granted statements, standardised tests, conceptions, and objects (which stand up to any imaginable trials) in a community.

Against this background, it is worth asking why Latour's claims have caused so much controversy. That is probably because Latour's understandings of the notions of 'fact' and 'nature' are different from the ones that have been generally accepted by most contemporary philosophers and natural scientists, i.e., the realist understanding of facts and nature as human-independent that make our propositions true or false. If a commentator is not careful in distinguishing different senses, s/he is easily led to a wrong track. In other words, Latour does not use 'fact', 'reality' and 'nature' as 'elevating words', as Ian Hacking has called their usage in philosophy (Hacking, 2001, pp. 22–23). 'Fact' for Latour is not a human-independent thing out there in the world, but quite a mundane entity, as we have seen. If we try to read Latour's texts with a realist metaphysical understanding of 'facts' in our minds, the situation becomes very confused. If facts are independent of us, it does not make good sense to talk of constructing facts or of different stages of fact-making. If a scientist complains that the story of the emergence of the hormone TRF shows how scientists discovered a fact, Latour is inclined to respond that this is not possible because 'TRF' and other 'facts' are by definition constructed. He does not want to question the solidity of facts (as collectively endorsed), but to show how, where and when they were created. (Cf. Latour and Woolgar, 1979, pp. 127 & 175)

We can now see that scientists and philosophers of science, on the one hand, and Latour, on the other hand, are at cross-purposes. From a certain metaphysical point of view, it is almost outrageous to claim that facts are not qualitatively different from fiction (Latour, 1987, p. 42), but this becomes understandable if one remembers that we are talking about collective stabilisations of statements. Any statement could, in principle, be questioned, but some achieve a foundational status, at least for a certain limited time. It is as if Latour's idea is to label different statements according to their status of general acceptability and stability. The difference between the statements of fact and fiction is a difference in degree.

It is thus clear that the realist understanding of 'fact' is not a subject for the construction talk. Interestingly, this is something that Latour too comments on. He says,

> Facts refuse to become sociologised. They seem able to return to their state of being 'out there' and thus to pass beyond the grasp of sociological analysis. In a similar way, our demonstration of the microprocessing of facts is likely to be a source of only temporary persuasion that facts are constructed. Readers, especially practising scientists, are unlikely to adopt this perspective for very long before returning to the notion that facts exist, and that it is their existence that required skilful revelation. (Latour and Woolgar, 1979, p. 175)

I think this observation hits the nail on the head and it is encouraging that Latour is able to see the different perspectives. Practising scientists or analytic philosophers do not normally talk of the stabilisation of assumptions when they write about facts. In actuality, Latour's non-transcendental 'facts' are not facts at all for the majority of contemporary scientists and philosophers; for them, they are what a community believes or takes to be a fact.

4 Latour as descriptivist

What I have said above represents an attempt to give a common-sense reading of Latour's work. The basic premise of this task is to take the anthropological perspective that limits all the analysis of science on the non-theoretical and non-transcendental surface level seriously. I believe we can continue this approach further and demystify a number of other claims and views of Latour.

When the reader comes across the idea that neuroendocrinologists are a "tribe of readers and writers" that spend most of their time reading and writing neuroendocrinological literature with various inscription devices found in their laboratory, one is taken aback. But we can now see that it makes good sense in Latour's mind-set. If an anthropological fieldworker enters a laboratory without preconceptions of their activity and their knowledge, and observes their activities from beginning to end, the observer will see that it does involve a lot of writing (Latour and Woolgar, 1979, pp. 56 & 69). Further, a major achievement for a research group is to produce a text which is published in a scientific journal and which will be read and cited by other scientists. Latour calls scientific writing 'fact-writing', which may precede many other activities but which aims at persuading others of the facticity of one's statements. The whole scientific activity seems to be directed towards this aim. As he vividly says, in order to persuade others that one's statements should be taken seriously, "rats had been bled and beheaded, frogs had been flayed, chemicals consumed, time spent, careers had been made or broken, and inscription devices has been manufactured and accumulated within the laboratory" (Latour and Woolgar, 1979, p. 88). In brief, if facts are understood as generally accepted statements, then one can say that scientists are fact-writers, because they try to make their statements as widely accepted as possible, and the best way to achieve this is to publish a paper in a highly esteemed scientific journal.

Latour's view of how objects in science are born and defined is very interesting. As with so many of his claims, they strike one as being radical at first sight, but turn out pretty trivial under a closer analysis. How are new objects born in the laboratory? By recording the answers that a new object "inscribes on the window of instruments" (Latour and Woolgar, 1979, p. 90). Or, "the new object is a list of written answers to trials" (Latour and Woolgar, 1979, p. 87). At the time of the emergence of an object, one cannot do much more than repeat the list of constitutive actions of the putative object in the following way: 'with A it does this, with C it does that', etc (Latour and Woolgar, 1979, p. 88). Further, Latour likes to talk of 'somethings' which do not first have names, but which receive names like 'somatostatin', 'polonium', 'anaerobic microbes', 'transfinite numbers', 'double helix', or 'Eagle computers' after the initial characterisations of their actions. This leads to a practice which presumes their existence independently of the trials that produced them (Latour and Woolgar, 1979, pp. 89 & 92).

It is noticeable that Latour refers to an underlying entity ('it does') as the source of the readings and observations that make the 'object'. One wonders whether this is a sign of the implicit acceptance of minimal realism in Latour. Naturally, Latour would not be willing to postulate a context-transcending entity, but it sounds as if he is nevertheless doing just that. I will come back to this question at the end of the paper. Second, Latour is clearly, and not totally unreasonably, describing the process that philosophers tend to call baptising or initiation ceremony of reference, made famous by Kripke (1980). Latour is clearly describing a baptismal of an unobservable object, and the reference of a term is fixed by a description of the causal powers that the object is supposed to have. There is nothing in these ideas that most contemporary philosophers of science would not be able to accept. Some of those who commit themselves to the Kripke-Putnam causal theory of reference accept now that description is needed at the time of baptism, but not required for successful reference determination thereafter.[5] However, according to Latour, if we add an item to the list of actions, we redefine the object (Latour and Woolgar, 1979, p. 88). This is exemplified in his comment on how TRF, an object of research of endocrinologists, is defined. A research of TRF in 1976 created a new object, which was "not the TRF

---
[5]Cf. Devitt and Sterelny, 1999, pp. 83–114.

of 1963, 1966, 1969, or 1975". Why is this the case? Because "from a strictly ethnographic point of view ...the object was constructed out of the *difference* between peaks on two curves" (Latour and Woolgar, 1979, pp. 125–126; original emphasis). At that time there was a change in the curves produced by a bioassay which were used to detect the presence of TRF. In other words, any and all the attributes associated with a putative object defines the object, or is part of the set of predicates which pick out a reference. A change in the object-characterising description changes the object or reference itself.

Latour observes perceptively that this way of defining objects raises a philosophical problem. If the object is defined by all the attributes given to it and observations made of it by scientists, can we say that it existed at all before the trials which were used to individuate the object? (We may add that this problem is additional to the one which changes the object whenever its description changes.) Latour answers his own question in the negative. He says it is wrong to speak of 'discovering microbes'. Instead, we should say that Pasteur constructed or "shaped" them (Latour, 1984, p. 88). Furthermore, 'microbe' "existed only for a time, in the absence of anything better, before it in turn was distorted" (Latour, 1984, p. 107). Isn't this an absurd statement? One thing is that Latour informs the reader that this is a conclusion reached from a practical and not from a theoretical point of view (Latour, 1984, p. 80), which is an allusion to his anthropological approach. But even if we adopt here a theoretical perspective, his view makes sense. *Strictly speaking*, Latour is correct. To the best of my knowledge, there is no object which would uniquely satisfy the description given by Pasteur. It is just as Latour says, subsequent scientists 'broke the microbe' into its constituents (Latour, 1984, p. 108). This is to say that Pasteur's 'microbe' refers to a number of separate entities and that the properties attributed by Pasteur probably do not hold of all those entities.

Philosophically speaking, Latour commits himself not only to a descriptive theory of reference fixing but also to what might be called a wide descriptive theory of reference. According to the wide descriptive theory of reference, in order for a term to refer, all the assumptions postulated about the putative reference have to be satisfied by the reference. This is a doctrine which has sometimes been used to explain Kuhn's idea of meaning change and specifically his claim that references change in theory transitions.[6] As Latour himself indicated, the doctrine is philosophically problematic. Most contemporary philosophers would not be willing to go as far as to claim that no microbe existed before Pasteur's postulations or that reference changes whenever theory changes.[7] Even Latour himself drafts a rudimentary an-

---

[6] Cf. Bird (2000, pp. 164–179); Kuukkanen (2008, pp. 62–63 & 188–189).
[7] One option is to adopt a causal theory of reference, already mentioned. Further, some

swer to the question of how to preserve the continuity of reference in theory transitions. He says that a matter of revealing the 'true agent' from the false ones requires showing that a new translation includes all the manifestations of the (true) earlier agent. Otherwise, the argument and scepticism about it continue and others will try redubbing (Latour, 1984, p. 81).

What this shows is that Latour comes across as more trivial and comprehensible than may look at first sight. Above all, he is not a metaphysical constructivist, but his construction talk derives from his anthropological point of view. The wide descriptivist perspective to reference is a consequence of this standpoint, as writing down all the characterisations of a putative reference without further theoretical reflection or critique makes the object identical with its description. Although these positions may not be philosophically favoured by most contemporary analytic philosophers, they nevertheless enable (one hopes) a rational dialogue with Latour.

5 Success and failure (un)explained

We have seen that symptomatic of Latour's anthropological approach is minimalism. It sticks to the notions that are immediately empirically verifiable and avoids using transcendental concepts. However, it is not clear whether it is able to offer us also explanations of science, or more specifically, of explanations of failures and successes in science. Further, in this context, it is worth noting that Latour has also formulated something that might well be called a theory of scientific advance. I will next attempt to evaluate how explanatory his model of science is with regard to success and failure.

It may appear that Latour's more positive philosophy of science is detached from his anthropological approach. This is however not the case. As we have seen, Latour demands that any concept used in explanations has to be empirically justified, and this is also the case with the 'theory' of science that arises out of Latour's other works and especially from *Science in Action*. In this sense, the deflationary and positive sides of Latour can be conceived of forming a continuum. My intention is to divert philosophers' attention away from the debate on social constructivism, where, in my view, there is less controversy to be found, to the questions where more significant disagreements may be expected to arise.

---

realistically minded philosophers, such as Philip Kitcher (1978), have suggested that we might be able to identify contexts where a reference can be taken to be properly referring from reference failures. Hartry Field (1973) has suggested that we could conceive of there being "partially denoted" entities, which may subsequently undergo "denotational refinement". For example, Newton's mass, according to Field, partially denoted relativistic mass and proper mass, and in the Einsteinian revolution went under refinement to denote only the latter.

It is true that not all Latour's books are as anthropological as *Laboratory Life*, which has served us as a kind of ideal model of the anthropological orientation. Nevertheless, the anthropological-empirical orientation characterises Latour's general approach to science also in his other works. For example, *Pandora's Hope*, published at the turn of 21st century, contains an anthropological study about soil research in the Amazon forest. Latour also confesses he believes in "a universalist anthropology", pointing out that also the "modernist settlement" makes an object of "a true anthropological curiosity" (Latour, 1999, p. 277; cf. p. 14), which also is the central topic in his book *We Have Never Been Modern* (1991). In a similar fashion as earlier works, Pandora's Hope is peppered with claims that the conclusions reached are empirically justified. The book talks about dropping modalities and fact-making (Latour, 1991, pp. 93–94), the impossibility of making any a priori divisions (e.g., pp. 86–87 & 126), empirical documentation of the historical conclusions (e.g., pp. 166–167), or how Latour simply follows "the veins and arteries" of science (p. 106). Most important, Latour's 'positive explanation' of science, his network model expounded in *Science in Action*, is alleged to be based on empirical findings.[8] His six general principles in *Science in Action* represent Latour's "personal summary of the empirical facts at hand after a decade of work in this area" (Latour, 1987, p. 17). Let us have a look at some of them.

For our purposes, the third and sixth principles are most relevant:

> *Third principle.* We are never confronted with science, technology and society, but with a gamut of weaker and stronger associations; thus understanding *what* facts and machines are is the same task as understanding *who* the people are. (Latour, 1987, p. 259)
>
> *Sixth principle.* The history of technoscience is in large part the history of all the little inventions made along the networks to accelerate the mobility of traces, or to enhance their faithfulness, combination and cohesion, so as to make action at a distance possible. (Latour, 1987, p. 259)

I do not intend to evaluate here how strongly these are supported empirically. In any case, although they represent a step towards abstraction and generalisation from what we find especially in *Laboratory Life*, Latour puts these forward as empirical hypotheses which ought to be debated and falsified if necessary.

---

[8]In a relatively recent book, *Reassembling the Social* (2005), Latour continues to explicate his 'actor-network-theory'. He attempts to redefine sociology not as the 'science of the social', but as a study of the 'tracing of associations' (p. 5). The 'movement of re-association and reassembling' is of specific interest and a problem that calls for a social explanation in this new 'sociology of associations' (pp. 7–9).

Further, Latour's model is meant not to provide an alternative epistemology, but to discard totally all cognitive explanations of science. However, interestingly, he proposed "a ten-year moratorium on cognitive explanations of science" promising to "turn to the mind" if anything remains to be explained after that (Latour and Woolgar, 1979, p. 280). This moratorium is also expressed in Latour's Rule 7:

> Before attributing any special quality to the mind or to the method of people let us examine first the many ways through which inscriptions are gathered, combined, tied together and sent back. Only if there is something unexplained once the networks have been studied shall we start to speak of cognitive factors. (Latour, 1987, p. 258)

Now it is more than twenty years since the moratorium was first time announced (in "Postscript to Second Edition" of *Laboratory Life*) and high time to update the situation. Should Latour keep his moratorium in place or is it time to employ cognitive explanations again?

As we just saw, central to Latour's explanation of science is networks. Latour does not make distinctions between science, technology and society. Outside and inside of science are linked to each other so that any sharp distinction between them is meaningless. According to Latour, there is a positive feedback loop between them: "the bigger, the harder, the purer science is inside, *the further outside other scientists have to go*" (Latour, 1987, p. 258, original emphasis). In other words, the existence of the 'inside' of science depends on the resources that scientists have managed to collect 'outside', i.e., what kind of funding, public image, political relations, etc. a particular institution or group has managed to gather. Latour takes all the people who provide the context of science as scientists as much as anyone who creates its content. Crucially, success in this 'outside' activity is dependent upon how many people are convinced that support and concentration on certain scientific work is "necessary for furthering *their own goals*" (Latour, 1987, pp. 156–157, original emphasis). A good example of how the content of science depends on the external enabling conditions is botany. Latour asks if botany may be constructed "everywhere in a universal and abstract space?" His answer is, "certainly not, because it needs thousands of carefully protected cases of dried, gathered, labelled plants; it also needs major institutions like Kew Gardens or the Jardin des Plantes where living specimens are germinated, cultivated and protected against cross-fertilisation...Botany is the *local knowledge* generated inside gathering institutions" (Latour, 1987, p. 229, original emphasis). The general lesson is that all sciences, no matter whether we are talking about the laws of physics, biology or mathematics, depend on the application of enabling material and non-material conditions (cf. Latour, 1987, p. 250).

'Technoscience' is thus entirely dependent on its ability to spread networks further. The explanatory model boils down to the struggle to extend one's own networks, persuade others to become its advocates and control their behaviour for this end. Latour introduces a term 'sociologics' (in contrast to logic), which focuses on the strength and number of associations in the network (Latour, 1987, p. 202). "The only things we want to know about these sociological pathways is where they lead to, how many people go along them with what sort of vehicles, and how easy they are to travel; not if they are wrong or right" (Latour, 1987, p. 205).

The question we have to ask is whether this model is enough to account for both successes and failures in science. Latour's seventh rule of methods instructs: "First to look at how the observers move in space and time, how the mobility, stability and combinality of inscriptions are enhanced, how the networks are extended, how all the informations are tied together in a cascade of re-representation" (Latour, 1987, pp. 246–247). The implication is that, if this is not enough, then we are allowed to look for cognitive explanations.

However, it should not be expected that the history of science provides us a certain correct model or shows indubitably that another model is incorrect. As John Preston has reasonably pointed out, it would be naïve to expect that the history of science provides us an unambiguous answer to the question whether the cumulativist or revolutionary model of scientific development is correct, for example. Up to a certain point, history is an interpretative and therefore philosophical discipline (Preston, 2008, p. 54). Therefore, the most sensible strategy is to focus on evaluating the explanatory power of rival models, and ask which one gives us the most satisfactory explanation.

On a certain level Latour's attitude is fully reasonable. He advises against using 'nature' (or 'God' for that matter) to explain why scientific disputes were settled (e.g., 94–95; 183). And this has to do with studying science in action. Scientific disputes are about what nature is or what is real, and it would be unreasonable to assume that one of the participants in these disputes possessed a God-like access to nature, truth or reality (cf. Bonjour, 1985, p. 7; Rescher, 1973, pp. 5–9). According to Latour, this would amount to Whig history which explains past developments by pointing out who was right and wrong. Latour remarks that one needs more fine-grained explanations, for example, as to why "people slowly turned N-rays into an artefact" (Latour, 1987, p. 100).

It is surely true that *at the time* physicists did not have the philosopher's stone which would have told what the true fact of the matter was. However, Latour also says that when the controversy is settled, nature is used as "the ultimate referee". Even if we were to agree with him that at the microsociological and -historical level the transcendental references cannot be used as explanatory notions, it is reasonable to ask whether avoiding using such notions leaves something unexplained when the results of the controversies are known. Let us therefore have a closer look at how Latour explains scientific advancement by studying his explanation of Pasteur's success.

As we have already seen, Latour adopts an agnostic stance. He makes no *a priori* distinctions between science and the rest of society, reason and force, main actors of the story, or in general, between 'allies' that make the pathways of science. However, he postulates that everything is involved in a relation of forces although he remains agnostic about the nature of these forces (Latour, 1984, pp. 7–9). According to Latour, what was striking in Pasteur's case is not his intellectual ingeniousness, but his ability to translate the intentions of the hygienists to support his own case so that both the hygienists and the Pasteurians were strengthened as a result. They together managed to make microbiology and the sanitization plans indisputable (Latour, 1984, pp. 34 & 54). According to Latour, this had nothing to do with Pasteur's cognitive capacities or with the purported truth of theories (in the modernist pre-settlement sense, at least). Pasteur was thus extraordinary only in "translating the wishes of practically all the social groups of the period, then getting those wishes to emanate from a body of pure research that did not even know it was applicable to or comprehensible by the very groups from which it came" (Latour, 1984, p. 72). For example, he had to convince others that not only medical practice but also laboratory work was relevant at the time when infectious diseases were killing people around them.

In other words, the advancement of science is making new allies and extending one's network further and further. Pasteur became undisputed; hygienists gained positions in public administration and replaced engineers. The hygienists thus used Pasteur to secure positions, which suited Pasteur very well (Latour, 1984, pp. 56–58). This also implies that there is no qualitative difference between 'right' and 'might' or 'real' and 'unreal'. Differences between these kinds of notions as well those between 'illogical' and 'logical', 'contradictory' and 'consistent', etc. are a matter of different strengths of forces in the network (cf. Latour, 1984, pp. 153–154, 155–157 & 183). Let us ask now yet more explicitly: How does Latour explain success and failure in science? Because of his sixth principle above, we already know that the history of technoscience is recording the nature and strength of the networks that makes science possible. And so, "every time a fact is verified and a machine runs, it means that the lab or shop conditions have been extended *in some way*" (Latour, 1987, p. 250; original emphasis). But what is the difference between the cases where science's predictions are fulfilled and those when they fail? Latour answers, "The rule of method to apply here is rather straightforward: every time you hear about a successful application of a science look for the progressive extension of a network. Every time you hear about a failure of science, look for what part of which network has been punctured. I bet you will always find it" (Latour, 1987, p. 249).

The explanation above offers us the following two principles: When there is success, networks extend. And when there is failure, networks shrink. But this strikes one as being no explanation at all, but more like an empirically based re-statement of the facts of the matter. It is not surprising that successful science has managed to develop more extensive support networks, including institutions, the support of funding bodies, practical applications and publicity. Isn't it part of the definition of why it is successful? One wonders why did the particular application of science become such in the first place. Is it just because the participants in its networks were more cunning negotiators? Perhaps the networks extend because the forces behind it have been stronger and better at negotiation and diverting others for their cause. As we just saw, Latour attributed such skills to Pasteur. And when the networks fail, they do so because the forces are weaker and less skilful. This admittedly is *an explanation*, but we have to ask whether it is good enough.

Would Latour's model explain satisfactorily, for example, the failure of Lysenko's biology in the Soviet Union, considering it had managed to build the most extensive networks with the most powerful 'outside' support structures? Lysenko's theory received the acceptance of the Bolsheviks and became the Soviets' official application of dialectical materialism after 1935, gaining the support of Stalin, which enabled exceptional theoretical and material resources for its application and implementation (Lecourt, 1977, p. 99). It became part of the new science of agronomy and the collectivisation of Soviet agriculture, which led to wide attempts to transform numerous plants into others and in some cases the utilisation of wide areas of land for this purpose. Further, these experiments received wide publicity as well, as in nearly every issue of *Agrobiologia* from 1950 to 1955 articles appeared reporting transformations of wheat into rye and vice versa, barley into oats, peas into vetch, vetch into lentils, cabbage into swedes, firs into pines, hazelnuts into hornbeams, alders into birches and sunflowers into strangle weed (Medvedev, 1969, p. 170).

The question we have to ask is why did Lysenko's biology fail despite all this? Why did the extension of his networks come to an end? The obvious reason that he lost his most powerful supporter, Stalin, cannot be right, because his success survived well into the 1960s and received Khrushchev's personal support still at the beginning of the 1960s (e.g., Medvedev, 1969, p. 205). To say that the Western genetics had yet more powerful networks and thereby managed to divert Lysenko's biology (or the advocates of it) for its benefit sounds ad hoc. At the time, contacts were not direct and multiple, but Lysenko's biology nevertheless came quickly to its end around the mid 1960s. One cannot either really say that Lysenko's opponents were better at negotiating, as he himself was obviously a master in that skill.

We recall that Latour makes no *a priori* assumptions, which applies also to the distinction between human and non-human actors in the networks. Would it not then be possible to say that the reason for the failure is that non-human 'actants,' as they are called in Latour's parlance, did not cooperate, or Lysenko didn't manage to persuade non-human agents to further his goals? Indeed, we might. But what would this type of explanation amount to? David Bloor remarks that both Mendelism and Lysenkoism were engaged with nature but in two different ways (Bloor, 1999, p. 88). The point is to spell out the difference in their engagements so that it would add something to the explanation of the asymmetrical outcomes of these traditions. The upshot is that, if Latour were to say that Lysenko did not deal with non-human actants (i.e., 'nature') in the correct way, it would raise cognitive questions of the correctness and justification of his theories and experimental predictions. One would be bound to ask whether there was something wrong with his engagement with non-human actants. Did he not ask the right questions or place the actants under the right kinds of tests with respect to the expected answers?

Lysenko's failure could, indeed, indicate that his ideas and predictions of these non-human actants were not correct, and this, in turn, would provide us a more complete explanation of the failure. Naturally, Latour can choose not to engage in any deeper examination of non-human actants and the relationship in which they stand to theories and experiments. But this would mean admitting that actants (human and non-human alike) do explain, or are at least part of the explanation of, the asymmetrical outcome between successful and unsuccessful networks without trying to uncover what lies beneath; without trying to specify the exact reasons for the outcome. Interestingly, Latour goes a long way towards conceding this conclusion: "Why can't we say that Pasteur was right and Pouchet was wrong? Well, we can say it, but only on the condition that we render very clearly and precisely the institutional mechanisms that are *still at work* to maintain the asymmetry between the two positions" (Latour, 1999, p. 168; original emphasis). Latour thus accepts that the question can be put in these terms although he maintains that the support networks of science would explain the asymmetrical position between these two traditions. But, surely, the actants other than 'institutional mechanism', including non-human, are also responsible for the failure. One wonders what they are and how their role can be used to explain the outcome. Would it not be conceivable to accept that one (if not the only) explanatory factor of Lysenko's failure is that there was something wrong with how Lysenko described nature? One possibility is to say that the extension and upholding the networks became gradually more and more laborious and difficult when faced with the apparent failures of his theories and applications. This would be to say that, ultimately, Lysenko's biology failed because it was too wrong despite a determined effort to the contrary. Similarly, one might insist that the failure of Pouchet to get non-human actants to work in his favour as well as Pasteur did is a symptom of being wrong about nature in some important way.

Whether the interpretation above is true or not, these *kinds of explanations* (that use cognitive notions) would form a more complete explanation than a mere referral to forces and agents that tried to extend Lysenko's networks. However, it is important to add a caveat here. I am not suggesting that Pasteur's success shows that he was necessarily right; or more generally, that the success of a scientific theory is an infallible sign of its truthfulness. Kuhn's *Structure of Scientific Revolutions* (1970), Laudan's *The Confutation of Convergent Realism* (1981) and the subsequent discussion on the progress of science have made the unfeasibility of this kind of reasoning painfully evident.[9] There have been numerous successful theories in the past, which are however judged to be false if measured on modern standards and conceptions. But what can be said is that the refusal to consider any cognitive explanations of science threatens to leave one's explanations half-baked. First, one needs to give some account of why some networks extend while others fail. Second, if one attributes an explanatory role symmetrically to all kinds of actants to explain the asymmetrical state of networks, then one has to be ready to consider that role more specifically, including of what can be said of the properties and causal powers of the entities postulated.

In the current situation, where the successes of certain scientific traditions are evident, a mere reference to the fact that some networks extend while other shrink, would very likely leave something unexplained. Although it is challenging to specify the link between success, failure and the truth in science, it is reasonable to suggest that the cognitive considerations in terms of rightness and wrongness of theories become more compelling when failures and successes of them and their advocates become more blatant. And although the predictive and explanatory success of certain theories is not necessarily a sign of them being true or approximately true, it can be.

6 The end of the moratorium

I wish now come to back to the question of social constructivism. My conclusion above was that Latour is not an ontological or metaphysical

---

[9]There is, of course, a plenty of discussion about the relationship between success/failure and the truth. Laudan (1981) is a central piece. For some other important initiatives on the topic, cf. Van Fraassen (1980), Lipton (2004) and Psillos (2005).

social constructivist despite the appearance of some of his statements. It is tempting to classify Latour's network model as social constructivist in the epistemological sense, because scientific knowledge in his model looks like a result of different kinds of forces furthering their interests and causes. However, this too would be a mistake, as we see soon. Further, one might be tempted to read Latour as saying that the transcendental layers of nature, world, reality etc. do not exist. However, this does not seem to be the correct interpretation in light of Latour's own words. Alternatively, it would show that Latour is internally inconsistent.

Latour does not deny that there is reality, nature or truth (now understood in the philosophers' transcendental sense). Latour says, "Philosophers fool themselves when they look for a correspondence between words and things as the ultimate standard of truth. There is truth and there is reality, but there is neither correspondence nor *adequatio*" (Latour, 1999, p. 64; cf. also p. 15). What he specifically denies is the modernist settlement, i.e., the view that one could have independent access to them or that one could separate their influence from many other factors which have a role to play in the construction of scientific knowledge. According to Latour, in science studies, it does not make sense to talk independently of epistemology, ontology, psychology, politics, or theology. The central point is that they all *"go hand in hand and are aiming at the same settlement"* (Latour, 1999, pp. 13–14; original emphasis). Latour actually purposefully mixes epistemology with ontology (e.g., Latour, 1999, pp. 93 & 141). And this attitude implies that natural and social factors cannot be separated and talked about independently, and 'forces' in action contain inseparably both human and non-human elements (e.g., Latour, 2005, pp. 254–255). In brief, they form a hybrid.[10] Latour writes, "science studies does not say that facts are socially constructed... There exists only *one* settlement, which connects the questions of ontology, epistemology, ethics, politics, and the technology. There is thus no longer much sense in pursuing in isolation questions like 'how can a mind know the world outside'" (Latour, 1999, p. 293; original emphasis).

Now the crucial question is whether we should accept this kind of 'nonmodernist' stance, which might be said to be a form of epistemological antirealism in the philosophy of science. The problem is not the non-existence of the natural world, but that it is impossible to have independent unmediated access to it. However, it seems that even Latour himself cannot stick to the

---

[10] It is worth pointing out that this is not true of another major tradition in the sociology of science, The Sociology of Scientific Knowledge, although one can find generalisations that suggest otherwise (cf. Sokal and Bricmont, 1998). Bloor quite explicitly commits to a view that both social and non-social factors have a role to play in the construction of scientific knowledge, and that their role is analysable (cf. Bloor, 1991, p. 166; Bloor, 1996, p. 84; Bloor, 1999, pp. 81, 88, 90, 93, & 102; similarly Barnes, 1974, p. 43; Barnes and Bloor, 1982, p. 33).

restrictions of his model that forbid using the modernist kinds of context-transcending notions in the explanations. Earlier in the paper, it seemed that Latour implied that context-transcending entities are the causal factors behind the observable features used to define objects ("with A it does this, with C it does that"). The same sentiment is also found in his allusion that reality is defined by the resistance of trials that objects meet, as it sounds reasonable to assume that it is the world or objects in the world 'out there' that are the sources of this resistance. And as argued above, the nature of non-cooperating and cooperating non-human actants may need to be taken into account in the explanations of science's successes and failures. This is what I think shows the limitations of Latour's explanatory model of science. Latour's network model may be useful in giving an account of how the inside content of science and its outside support network are intertwined, but his explanation of science is insufficient at best. As much as any proper explanation of Lysenko's failure in biology may require a reference to context-transcending notions, so may explanations of success stories in science. At the very least, one has to be open for this possibility. All in all, Latour's examination of the history of science shows some signs of implicit inclination towards such context-transcending explanations as limiting cases of other explanatory strategies.

In conclusion, we have seen that conceptually it is possible to accept Latour's story about construction of facts, reality and scientific objects, and still maintain that there are facts, reality and scientific objects (although not necessarily in all cases) in the philosophers' transcendental sense. In judging Latour, my suggestion is not to focus on Latour's seeming ontological radicality or his apparent contradiction with common-sense approaches in philosophy, but to ask about the concrete value of his explanatory model of science. More specifically, his denial that 'nature' or 'reality' cannot be understood independently of social, political and technological aspects has to be examined critically. On several occasions, Latour remarks that the retrospective characterisation of the microprocesses in science often uses epistemological notions. The proper reaction is to point out that this is a laudable orientation, if we want to find a properly explanatory account of science at some future point of time. And if so, Latour's moratorium must be ended belatedly and science studies should consider employing cognitive explanations once again.

Bibliography

Barnes, B. (1974). *Scientific knowledge and sociological theory*. Routledge and Kegan Paul, London.

Barnes, B. and Bloor, D. (1982). Relativism, rationalism and the sociology of knowledge. In Hollis, M. and Lukes, S., editors, *Rationality and Relativism*, pages 21–47, Oxford. Blackwell.

Bird, A. (2000). *Thomas Kuhn*. Acumen, Chesham.

Bloor, D. (1991). *Knowledge and social imagery*. University of Chicago Press, Chicago, IL. 2nd ed.

Bloor, D. (1996). Idealism and the sociology of knowledge. *Social Studies of Science*, 26:839–856.

Bloor, D. (1999). Anti-Latour. *Studies in History and Philosophy of Science*, 30:81–112.

Boghossian, P. (2001). What is social construction. Times Literary Supplement, 13 February 2001.

Bonjour, L. (1985). *The structure of empirical knowledge*. Harvard University Press, Cambridge, MA.

Callon, M. and Latour, B. (1992). Don't throw the baby out with the Bath school! A reply to Collins and Yearley. In Pickering, A., editor, *Science as practice and culture*, pages 343–377, Chicago, IL. Chicago University Press.

Collins, H. and Yearley, S. (1992). Epistemological chicken. In Pickering, A., editor, *Science as practice and culture*, pages 301–327, Chicago, IL. Chicago University Press.

Devitt, M. and Sterelny, K. (1999). *Language and reality: An introduction to the philosophy of language*. Blackwell, Oxford. 2nd ed.

Field, H. (1973). Theory change and the indeterminacy of reference. *Journal of Philosophy*, 70:462–481.

Golinski, J. (1988). *Making natural knowledge. Constructivism and the history of science*. Cambridge University Press, Cambridge.

Hacking, I. (2001). *Social construction of what?* Harvard University Press, Cambridge, MA.

Kitcher, P. (1978). Theories, theorists and theoretical change. *Philosophical Review*, 87(4):519–547.

Kripke, S. (1980). *Naming and necessity*. Harvard University Press, Cambridge, MA.

Kuhn, T. (1970). *The structure of scientific revolutions*. University of Chicago Press, Chicago, IL. 2nd enlarged ed.

Kukla, A. (2000). *Social constructivism and the philosophy of science*. Routledge, London.

Kuukkanen, J.-M. (2008). *Meaning changes. A study of Thomas Kuhn's philosophy*. VDM Verlag Dr Müller, Saarbrücken.

Latour, B. (1984). *Les Microbes. Guerre et paix, suivi de Irréductions*. Éditions Anne-Marie Métailié, Paris. Page numbers refer to the English translation (Latour, 1988).

Latour, B. (1987). *Science in action: How to follow scientists and engineers through society*. Harvard University Press, Cambridge, MA.

Latour, B. (1988). *The pasteurization of France*. Harvard University Press, Cambridge, MA.

Latour, B. (1991). *Nous n'avons jamais été modernes. Essai d'anthropologie symétrique*. La Découverte, Paris. Page numbers refer to the English translation (Latour, 2006).

Latour, B. (1999). *Pandora's hope. Essays on the reality of science studies*. Harvard University Press, Cambridge, MA.

Latour, B. (2005). *Reassembling the social. An introduction to actor-network-theory*. Oxford University Press, Oxford.

Latour, B. (2006). *We have never been modern*. Harvard University Press, Cambridge, MA.

Latour, B. and Woolgar, S. (1979). *Laboratory life. The social construction of scientific facts*. Sage, London.

Laudan, L. (1981). The confutation of convergent realism. *Philosophy of Science*, 48:19–49.

Lecourt, D. (1977). *Proletarian science? The case of Lysenko*. NLB, London. Translated by Ben Brewster.

Lipton, P. (2004). *Inference to the best explanation*. Routledge, London. 2nd ed.

Medvedev, Z. A. (1969). *The rise and fall of T. D. Lysenko*. Columbia University Press, New York.

Newton, R. G. (2000). *The truth of science. Physical theories and reality.* Harvard University Press, Cambridge, MA.

Pickering, A., editor (1992). *Science as practice and culture.* Chicago University Press, Chicago, IL.

Preston, J. (2008). *Kuhn's* The structure of scientific revolutions. Continuum, London.

Psillos, S. (2005). *Scientific realism: How science tracks truth.* Routledge, London.

Rescher, N. (1973). *The coherence theory of truth.* Oxford University Press, Oxford.

Sokal, A. and Bricmont, J. (1998). *Fashionable nonsense.* Picador, New York.

Van Fraassen, B. (1980). *The scientific image.* Clarendon Press, Oxford.

François, K., Löwe, B., Müller, T., Van Kerkhove, B., editors,
Foundations of the Formal Sciences VII
Bringing together Philosophy and Sociology of Science

Albert Lautman: Dialectics in mathematics

BRENDAN LARVOR*

Department of Philosophy, University of Hertfordshire, Hatfield, Hertfordshire AL10 9AB, United Kingdom
E-mail: b.p.larvor@herts.ac.uk

Albert Lautman (1908–1944) is a rare example of a twentieth-century philosopher whose engagement with contemporary mathematics goes beyond the 'foundational' areas of mathematical logic and set theory. He insists that (what were in his day) the new mathematics of topology, abstract algebra, class field theory and analytic number theory have a philosophical significance that distinguishes them from the mathematics of earlier eras. Specifically, these new areas of mathematics reveal underlying dialectical structures not found in earlier mathematics. In a series of short papers and two longer theses (*Essay on the unity of the mathematical sciences in their current development* and *Essay on the notions of structure and existence in mathematics*)[1], Lautman argues this claim from a philosophical perspective rooted in certain of the later dialogues of Plato. However, Lautman was not satisfied with Plato's conception of the relation between dialectical Ideas and the matter in which they are realised. In one of his last papers, *New research on the dialectical structure of mathematics*[2], Lautman bolsters his Platonism with an appeal to Heidegger's 'ontological' distinction between phenomenology and science.[3] We may therefore regard this paper as the most advanced expression available of Lautman's philosophy of mathematics.

---

*A French version of this paper was published as *Albert Lautman, ou la dialectique dans les mathématiques* in the journal *Philosophiques* 37(1):75–94, 2010. I am grateful to David Corfield, Nicholas Joll and Danièle Moyal-Sharrock for their careful reading of earlier versions of this paper.

[1] Henceforth, page numbers refer to the 2006 Vrin edition of Lautman's complete works, (Lautman, 2006).

[2] (Lautman, 2006, pp. 235–257); this paper was first published in 1939 in a series edited by Jean Cavaillès and Raymond Aron.

[3] As expressed in Heidegger's 1928 lecture *Vom Wesen des Grundes*. Quotations here are from McNeill's 1998 translation *On the Essence of Ground*. Lautman quotes Corbin's 1938 French translation.

Received by the editors: 17 December 2008.
Accepted for publication: 25 March 2010.

In this paper, I shall first explore Lautman's conception of dialectics by a consideration of his references to Plato and Heidegger. I shall then compare the dialectical structures that he found in contemporary mathematics with the model that emerges from his philosophical sources. I shall argue that the structures that he discovered in mathematics are richer than his Platonist model suggests, and that Heidegger's 'ontological' distinction is less useful than Lautman seemed to believe.

1 Plato

In his major case studies, Lautman developed a picture of modern mathematics (that is, mathematics in the early twentieth century) as the expression or realisation of fundamental conceptual oppositions (such as continuous/discontinuous, global/local, finite/infinite, symmetric/anti-symmetric).[4] He referred to the opposing terms as *notions*; dialectical *Ideas* envisage possible relations between such pairs of dialectical notions (Lautman, 2006, pp. 242–243). This terminology is a conscious reference to Plato, and he is careful to distinguish his appeal to Plato from 'Platonism' as philosophers of mathematics usually use the term. In philosophy of mathematics, 'Platonism' usually denotes the view that mathematical objects exist independently of the thought and talk of mathematicians. Lautman insisted that this was a misreading of Plato (Lautman, 2006, p. 230); in any case, this kind of 'Platonism' is not Lautman's view.[5]

Lautman never quotes Plato directly, and he mentions just three Platonic texts: *Philebus*, the *Sophist* (twice), and *Timaeus* (twice). Scholars usually count these among the 'later' dialogues of Plato (though the *Sophist* is continuous with the *Theaetetus* and implicitly refers to the *Parmenides*— both middle period dialogues). What matters for our purpose is that Plato's *theory of forms* is largely absent from his later works. The *Ideas* in the later dialogues are not blueprints for material objects. Similarly, Lautman's mathematical Platonism was not a 'copy-theory'. As he points out, we might think of material reality as inchoate matter somehow shaped into material copies of non-material 'forms', but this model cannot apply to the relation between mathematical theories and the dialectical ideas that (in Lautman's term) *dominate* them (Lautman, 2006, p. 238).

---

[4]This list is drawn from the two long essays. In *New Research on the dialectical structure of mathematics* he offers a slightly different list of dialectical pairs, "wholes and parts, situational and intrinsic properties, basic domains and objects defined on these domains, formal systems and their models, etc.." (Lautman, 2006, p. 243)

[5]"Dans le débat ouvert entre formalistes et intuitionnistes, [...] les mathématiciens ont pris l'habitude de désigner sommairement sous le nom de platonisme toute philosophie pour laquelle l'existence d'un être mathématique est tenue pour assuré [...] c'est là une connaissance superficielle du platonisme [...]" (Lautman, 2006, p. 230).

1.1 The *Sophist*

In a short paper of 1937 called *L'axiomatique et la méthode de division*[6], Lautman refers to *Philebus* and the *Sophist* together:

> The movement from so-called 'elementary' notions to abstract notions does not [...] appear as the subsumption of the particular under the general, but rather as the division or analysis of a 'mixture' which tends to yield simple notions in which this mixture participates. It is, therefore, not the Aristotelian logic of genus and species at work here, but the Platonic method of division, as taught in the *Sophist* and *Philebus*, in which the unity of Being is a unity of composition and a starting-point in the search for principles that are unified in Ideas.[7]

The *Sophist* is a discussion between a young man, Theaetetus, and a stranger from Elea, "a comrade of the circle of Parmenides and Zeno, and a man very much a philosopher" (216A). The initial question is whether the words 'sophist', 'statesman' and 'philosopher' name one, two or three types of thing, and what that thing is or those things are. The nameless stranger asks for an "interlocutor [who] submits to guidance easily" (217D); Socrates proposes young Theaetetus. Thus, Plato allows the unnamed philosopher to develop his position at length without having to fend off a Socratic interrogation (this is a feature of Plato's later works; in the eponymous dialogue, Timaeus has the floor to himself after the preliminary civilities). Thereafter, Socrates vanishes from the text, so we do not have the luxury of inferring Plato's view from Socrates's words.

The Eleatic philosopher proceeds by division, that is, by making one distinction after another. He illustrates this technique with the term 'angler'. He first distinguishes gathering arts from manufacturing arts; then the gathering arts are divided into trading and 'mastering' or getting the better of; getting the better of divides into competition and hunting; hunting divides according to quarry (animal or other); animals swim or walk; swimming animals divide into water-fowl and fish; fishing divides into trapping (with nets, traps, etc.) and striking; striking divides into striking down with a trident and up with a hook. The resulting tree of categories is his account of 'angler'. He then proceeds to apply the same technique to the term

---

[6] *Axiomatics and the method of division*; (Lautman, 2006, pp. 69–80).

[7] "Le passage des notions dites 'élémentaires' aux notions abstraits ne se présente donc pas comme une subsumption du particulier sous le général mais comme la division ou l'analyse d'un 'mixte' qui tend à dégager les notions simples auxquelles ce mixte participe. Ce n'est donc pas la logique aristotélicienne, celle des genres et des espèces qui intervient ici, mais la méthode platonicienne de division, telle que l'enseignent le Sophiste et le Philèbe pour laquelle l'unité de l'Être est une unité de composition et un point de départ vers la recherche des principes qui s'unissent dans les Idées" (Lautman, 2006, pp. 78–79).

'sophist', and this discussion occupies the remainder of the dialogue. The Eleatic philosopher develops several different accounts of 'sophist' (231D-E), which leads to a methodological discussion, including a debate about the possibility of numbering non-beings (238B). The discussion refers to itself, because Theodorus introduced the Eleatic stranger as a philosopher, presumably in virtue of his logical technique (253C).[8] But if the method of division turns out to be merely a spurious word-game, then perhaps *he* is a sophist. Certainly, his choices of divided categories seem arbitrary. For example, he might have divided fishing according to whether or not bait is used, in which case trident-fishing and net-fishing would have been divided from angling and the use of baited traps. Young Theaetetus submits to the philosopher's guidance rather too easily, and certainly more easily than Socrates would have done.

Whatever Plato's intent in giving an unnamed, generic Eleatic philosopher an easy ride, Lautman takes the method of division as an unproblematic technique, and makes no mention of its proper companion, the 'method of collection'. In the text immediately before the excerpt quoted above, Lautman runs through a list of mixtures, that is, mathematical items that 'participate' in two heterogeneous categories. Namely: arithmetical equality is the only equivalence relation such that the number of equivalence classes equals the cardinality of the base domain; the idea of multiplication refers both to the creation of arithmetical products and to the idea of operators on a domain; unity can be thought of either as the unit element of a ring of numbers or as the identity element in a domain of operators; the length of a segment depends on the size of the segment but at the same time depends on a convention; absolute value in classical algebra includes the notion of ordering but also the notion of the completeness of a field. He goes on to claim that some of these mixtures (arithmetical equality; multiplication; absolute value) are examples of the dialectical relation between the intrinsic and relational properties of mathematical objects (Lautman, 2006, pp. 78–79). He then suggests that, "the distinction thus established at the heart of a single concept between the intrinsic properties of an object [...] and its potential for action [on other objects] seems to resemble the Platonic distinction between the Same and the Other [...]"[9]. For Lautman, then, these mathematical items (equality, multiplication, unity, length and absolute value) all have, in some sense, one foot in each of two camps. We shall see this pat-

---

[8]But cf. Trevaskis's argument that there is more to the philosopher's technique than the method of division (Trevaskis, 1967).

[9]"La distinction qui s'ètablit ainsi au sein d'une même notion entre les propriétés intrinsèques d'un être ou d'une notion et ses possibilités d'action nous semble s'apparenter à la distinction platonicienne du Même et de l'Autre qui se retrouvent dans l'unité de l'être" (Lautman, 2006, p. 79). (Translation note: this translation is a little free in order to preserve Lautman's special sense of *notion*).

tern again in the fifth chapter of the *Essay on the notions of structure and existence in mathematics*, in which Lautman explores another collection of mathematical 'mixtures'. Notice, though, that the pairs of notions in this list are not pairs of conceptual opposites. He has this in common with the Eleatic philosopher; swimming is not the opposite of walking, nor is fish the opposite of fowl. The fact that these pairs are *not* conceptual opposites raises the question why the Eleatic philosopher divides categories into pairs (rather than triples, quadruples, *etc.*), with all the resulting awkwardness and arbitrariness. In another late work that Lautman mentions, *Timaeus*, Plato divides living creatures into four classes according to habitat: gods in heaven, birds in the air, land animals and water animals (39–40). Similarly in *Philebus*, when Socrates describes the method of division he requires only that a category be divided into a finite number of sub-categories (16D). The view that dialectics relates notions in pairs is indeed present in the *Sophist*, but only in the figure of the generic Eleatic philosopher. It does not seem to have been Plato's doctrine.

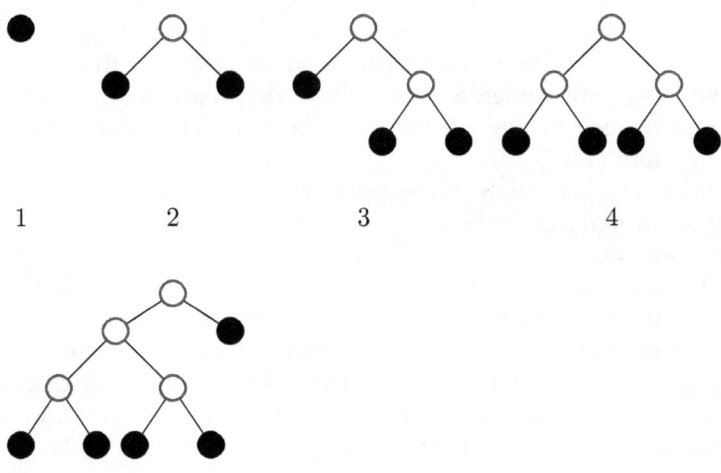

FIGURE 1. Lautman's Tree-diagrams.

The final reference to the *Sophist* is rather indirect. Lautman ends the *Essay on the notions of structure and existence in mathematics* with a gesture towards the thought that there is a developmental order among dialectical Ideas and from Ideas to mathematics. Lautman appeals to the works of Oscar Becker and Julius Stenzel on number in Plato and Aristotle. Lautman supplies a diagram taken from Stenzel (cf. Figure 1). Iterations of the Ideas 'one' and 'pair' produce 'Idea-numbers' (represented by the tree-diagrams), which in turn engender arithmetical numbers (represented by the black dots). Lautman's discussion is confused and inconclusive. He reproduces this diagram in the main text *and* in the footnotes, and mentions some reservations on Becker's part without discussing them (Lautman, 2006, p. 230, Figure 9). In any case, as Lautman acknowledges, Becker and Stenzel were both reading Plato through Aristotle (this was also Heidegger's procedure).[10] Having made this gesture, Lautman then turns to the relationship between mathematics and physics. The brief, inconclusive discussion with its pointlessly repeated diagram suggests some haste and dissatisfaction on Lautman's part.

1.2 *Philebus*

Lautman mentions *Philebus* once, in the quotation given above, as a source for the method of division. In this dialogue, Philebus, one of Socrates's young companions, holds that the good for man is pleasure. Socrates sets out to contest this, and to argue that intelligence (including knowledge and judgement) is better than pleasure. Before proceeding to his argument, Socrates makes a methodological digression. He describes the method of division (16C–17A), and insists that scientific understanding of a topic requires knowledge of the structure generated by successive distinctions. Unlike the Eleatic philosopher, Socrates allows that a category may divide into more than two subclasses; all he insists is that the number of subclasses should be finite (16D).

Early in the dialogue, Socrates points out that a life of pure intellectual activity is not suitable for men (21E). The good life for men must include some sensuous enjoyment as well as intellectual activity—though this being Plato, the intellectual side has priority. The crucial point is that the good for men is a mixture of heterogeneous elements (sensual and intellectual). This would present a paradox, if the method of division had the Aristotelian purpose of establishing a taxonomy. In the 'Aristotelian logic of genus and species', an object that seems to belong to two different species would be a counterexample to the taxonomy (as, for example, the duck-billed platypus threatens the category 'mammal'). In contrast, a Platonic system of Ideas is somehow prior to and independent of the objects that participate in those Ideas. An object can participate in more than one Idea (for example, a physical object might be both red and round). This mixing of Ideas occurs in other later dialogues. It is one of the principal explanatory motifs in the physics of *Timaeus* (34–35, 59–61), and we have already seen it in the *Sophist*.

For our purposes, the significant outcome of *Philebus* is that every human life must embody a mixture of sensual and intellectual goods. Pre-

---

[10] Heidegger (1925, p. 8). Lautman was not the only one to find Stenzel's reading of Plato on arithmetic more suggestive than clear. Cf. Cornford (1924); Shorey (1924).

cisely which goods and how they connect will vary from life to life. Perhaps someone's enjoyment of wine will develop an intellectual aspect as connoisseurship. To take a different example, intellectual work may offer some pleasures and satisfactions (though this is not, in Plato, the reason why it is good). One might imagine a life in which intelligence and pleasure were entirely separate departments, though it is hard to imagine desiring such an existence. Every human life will embody this dialectic in some way, and on philosophical examination will disclose it. Lives that lack one or other element must show that lack as an inadequacy or discontent. Indeed, we would need this dialectic of the sensual and the intellectual in order to understand the distempers and changes within a particular life. The background dialectical structure explains why a life given excessively to either sensuous pleasures or intellectual goods would be unsatisfactory.

Philebus, then, gives us an ethical analogy of Lautman's account of dialectical Ideas in mathematics. In Lautman's terminology, pleasure and intelligence are 'notions' and the possibility of relations between them is an 'Idea'. This dialectical structure does not specify which pleasures and thoughts will actually obtain. As Lautman says of mathematical Ideas, "As they are merely sketches of eventual positions, [Ideas] do not necessarily entail the existence of particular beings capable of sustaining the relations that the Ideas outline".[11] These notions come into relation through the interplay of particular thoughts and pleasures, and there is no predicting the detail of that interplay from the bare dialectical structure. The wine connoisseur's knowledge inflects his pleasure in boozing. Pleasure and intelligence relate quite differently (but no less intimately) in the rare but precious moments of insight in the work of a scientist. Similarly, Lautman maintains there is an indefinite variety of ways in which any dialectical relation between notions might manifest itself in actual mathematics, and it is not the business of philosophers to attempt to predict or circumscribe these relations (Lautman, 2006, p. 229).

1.3 *Timaeus*

Lautman's two references to *Timaeus* (Lautman, 2006, pp. 231 & 267) both remind us that for Plato, the creation of a material world is possible only if there is already a 'geometrically ordered receptacle' called 'place'.[12] Crucially, *different* objects may (at different times) occupy the *same* place. Thus, 'place' depends for its intelligibility on an anterior dialectical pair:

---

[11]"Étant seulement dessin de positions éventuelles, elles n'entraînent pas forcément l'existence d'êtres susceptibles de soutenir entre eux les relations qu'elles ébauchent" (Lautman, 2006, p. 243).

[12]"[...] le réceptacle d'une qualification géométrique" (Lautman, 2006, p. 231); translation note: the more literal "receptacle of a geometric qualification" makes little sense); "le lieu" (Lautman, 2006, p. 267). Cf. *Timaeus* §§48–9.

same/other. As we saw above, Lautman regards 'the distinction [...] between the intrinsic properties of an object [...] and its potential for action' on other objects as an expression of the same/other relation. In both cases, reference to Timaeus enables Lautman to shift from philosophy of mathematics to philosophy of physics. Lautman argues that the natural world is mathematically intelligible because the same dialectical structures underlie both physics and mathematics. He offers enantiomorphic crystals as an example of a physical phenomenon in which dialectical opposites (in this case, symmetry and dissymmetry) are 'mixed'. (This paper will not further discuss Lautman's philosophy of physics.) Here, as in his allusion to Stenzel's work on number, Lautman is trying to illustrate his thought that the intelligibility of mathematics and physics requires a prior dialectical order. In both cases, his exposition stumbles over Plato's inability to say what 'dialectical priority' means.

For Lautman, then, the method of division reveals dialectical 'notions' (in his special sense of the word), and with them the Ideas of relations between these notions. However, Lautman does not offer sequences of distinctions. His notions do not form tree-shaped accounts like those of the philosopher in the *Sophist*. As the quotation at the head of this section suggests, what he takes from these later dialogues is the thought that a particular can participate in heterogeneous categories simultaneously. In some of his examples, the notions 'mixed' in a mathematical theory are merely different (such as ordinal and closure), while in others they are opposites (as in the cases where he sees mixtures of finite and infinite mathematics).

2 Plato does not suffice

Lautman scattered references to Plato throughout his works; Heidegger, on the other hand, does not feature anywhere in his writing other than the discussion in *New research on the dialectical structure of mathematics* and implicitly in some brief remarks in the conclusion to *Essay on the notions of structure and existence* (Lautman, 2006, pp. 228–229).[13] We may therefore suppose that Lautman turned to Heidegger in order to solve a particular problem in his overall Platonism.[14] Moreover, the Heideggerian text that he refers to, *On the Essence of Ground*, is a meditation on the 'ontological difference' between the 'ontic' concepts employed in the sciences and the

---

[13] The sole exception is in a short piece of 1933 *Considérations sur la logique mathématique*. But here he discusses the use that the intuitionists made of phenomenology and makes no commitment of his own: "Les intuitionnistes se rattachent par là aux phénoménologues disciples de Husserl, Heidegger, et Oscar Becker" (Lautman, 2006, p. 43).

[14] Which is not to suggest that Lautman chose Heidegger arbitrarily, given his references to Plato; Heidegger prefaced *Being and Time* with a quotation from the *Sophist* (244a), and he devoted his lectures of 1924/25 to that same dialogue.

underlying 'ontological' concepts disclosed by phenomenology. The relation between dialectics and mathematics was clearly problematic for Lautman. On one hand, he was committed to his Platonist view that Ideas are somehow prior to the matter that they dominate, and which participates in them. In a talk given in 1937, Lautman claims that, "The reality inherent in mathematical theories is due to their participation in an ideal reality which dominates mathematics, but which cannot be known except through mathematics"[15]. He knew that the logical empiricist mainstream would regard his view as a mystification, "as obscure as the mystical beliefs of primitives in the participation of subjects in objects of which Mr. Lévy-Bruhl speaks".[16] He retorts that, on the contrary, empiricism (whether Aristotelian or Viennese) separates thought from experience and thus makes a mystery of the fact that we find nature mathematically intelligible. Moreover, a tautological view of mathematics separates the discovery of truth from the quest for reality (since tautologies do not require reference to any reality). Empiricism, he thought, deprives science of its spiritual dignity and value. Thus, it is scientifically and spiritually vital to insist on the reality of dialectical notions and the Ideas of their possible relations prior to their realisation in particular cases. On the other hand, notions only come into relations with each other when 'mixed' in particulars. Towards the end of the *Essay on the notions of Structure and Existence*, Lautman characterises Ideas of possible relations between notions as 'problems' or 'questions' and actual (realised) relations between notions as 'logical schemas':

> The logical schemas that we have described are not prior to their realisation at the heart of a theory; what is lacking from [...] the extra-mathematical intuition of the urgency of a logical problem is that it must have material to dominate, for the idea of possible relations to give birth to a scheme of real relations.[17]

Before the development of the mathematical theory that solves the problem, there is only "the experience of the urgency of problems".[18] However, this formulation makes it sound as if we are concerned with the psychology

---

[15]"La réalité inhérente aux théories mathématiques leur vient de ce qu'elles participent à une réalité idéale qui est dominatrice par rapport à la mathématique, mais qui n'est connaissable qu'à travers elle" (Lautman, 2006, pp. 67–68).

[16]"[...] aussi obscures que les croyances mystiques à la participation du sujet à l'objet chez les primitifs dont parle M. Lévy-Bruhl" (Lautman, 2006, p. 64).

[17]"Les schémas logiques que nous avons décrits ne sont pas antérieurs à leur réalisation au sein d'une théorie; il manque en effet à ce que nous appelons plus haut l'intuition extra-mathématique de l'urgence d'un problème logique, une matière à dominer pour que l'idée de relations possibles donne naissance au schéma de relations véritables" (Lautman, 2006, p. 229).

[18]"Le seul élément *a priori* que nous concevions est donné dans l'expérience de cette urgence des problèmes[...]" (Lautman, 2006, p. 229).

of mathematicians. That is not what Lautman had in mind. In the introduction to the *Essay on structure*, he concludes a discussion of Hilbert and Brunschvicg by insisting that, "Between logical deduction and the psychology of the mathematician, there must be space for an intrinsic characterisation of reality".[19]

What he requires, then, is a philosophical idiom in which this feeling for the urgency of a problem is more than a *mere* feeling. For, if this sense of urgency is no more than a psychological urge, then its content cannot have the logical significance that Lautman's Platonism requires. At the same time, this 'extra-mathematical intuition of the urgency of a logical problem' cannot be a mysterious sensitivity to a world of Ideas that exist prior to the activity of mathematicians. We have already seen that he rejects that kind of naïve Platonism. It is to fill this need that Lautman turns to Heidegger.

3 Heidegger

Lautman was familiar with *Being and Time*,[20] but he appeals to a much shorter work of Heidegger's: *On the Essence of Ground* (1928).[21] In this lecture, Heidegger tries to clarify his distinction between 'ontological' and 'ontic' (or equivalently, between 'being' and 'beings'). He does this through a meditation on the history of philosophy that resists summary, but the central thought is as follows. From Kant, we learn that metaphysics (or in Heidegger's terminology, 'ontology') is not directly concerned with what the world is like 'in itself'. Rather, ontology primarily reveals the deep structure of how we go at the world and go on in it. In this exposition, 'the world' should be read in something like the sense it has when one says that people who routinely practice sympathetic magic 'live in a different world' from people who routinely pursue their ends by modern scientific means. In this sense, only humans live in 'the world'. Cats and dogs occupy physical space, but they do not have a deep and largely inarticulate sense of what the world is like and how it works that shapes and guides their activities. The crucial feature of humans is that we enquire. Our questions may be practical (such as "why is my knee throbbing?" or "will I have to mix some more cement to finish this wall?") or they may be part of an advanced science. Scientific or not, every question has built in some assumptions about the form of the answer. A sleeper woken by a noise might ask, "what's that?" or "who's there?" or "did I imagine that?", depending on her expectations and cast of mind. This is true even of questions that seem to make no assumptions, such as "why is there something rather than nothing?" We did not have to ask

---

[19]"Entre la psychologie du mathématicien et la déduction logique, il doit y avoir place pour une caractérisation intrinsèque du réel" (Lautman, 2006, p. 129).

[20]Cf. his reference to it, (Lautman, 2006, p. 240).

[21]I am grateful to Nicholas Joll for his unstinting help with my reading of Heidegger.

about *things*. We might have asked, "Why is there stuff rather than void?" Scientific disciplines have their characteristic ways of going at the world: the modern physicist asks questions in the language of mathematics. This would be unintelligible to earlier students of nature who 'lived in another world' in the phenomenological sense (even if they understood the mathematics). Thus, our questions always reveal something of the deep structural features of the world as we take it.

What then of the 'ontological difference'? In a passage that Lautman quotes, Heidegger declares:

> The prior determination of the being (what-being and how-being) of nature in general is anchored in the "fundamental concepts" [*Grundbegriffe*] of the relevant science. In such concepts, space, place, time, motion, mass, force, and velocity are delimited, for example, and yet the essence of time or motion does not become an explicit problem. [...] The fundamental concepts of contemporary science neither contain the "proper" ontological concepts of the being of those beings concerned, nor can such concepts be attained merely through a "suitable" extension of these fundamental concepts.[22]

In other words, even the most fundamental concepts that scientists use are merely ontic. The corresponding ontological concepts lie outside the conceptual resources of science. Lautman insisted that the same is true of the dialectical notions and ideas that he discerns at work within mathematics. "Dialectic", he says, "is not part of mathematics, and its notions have no connection with the primitive notions of a theory".[23] Rather, dialectic is ontologically prior to mathematics in Heidegger's sense of 'ontological'. In the conclusion of *Essay on the notions of structure and existence in mathematics*, Lautman describes his philosophy of mathematics as phenomenological enquiry into the extra-mathematical intuition of or concern (*souci*) with the 'urgency' of a logical problem. We should read *souci* here as *Sorge* (concern or care) in the Heideggerian sense.

Heidegger's phenomenology is primarily concerned with our habits of mind and expectations as revealed in our questions (or rather, in our practices of enquiry). However, this enquiry into the structure of our active, questioning subjectivity also reveals the order of objective reality. Heidegger takes from Kant the thought that the coherent order of human subjectivity and the coherent order of the world are two sides of the same fact. Somehow, at an inexpressibly deep level, three aspects come together: the structure of our subjectivity, our busy activity using things to work on other things and the deep structures that we find embodied in the world.

---

[22] Heidegger (1967, p. 104–105). Quoted by Lautman (2006, p. 241).

[23] "La dialectique ne fait pas partie des mathématiques, et ses notions sont sans rapport avec les notions primitives d'une théorie" (Lautman, 2006, p. 242).

Lautman's Heideggerian account of the objectivity of mathematics seems to come to this: mathematical theories 'participate' in dialectical Ideas, in the sense that they relate dialectical notions. The Ideas pose vague and nebulous questions (in that they suggest the possibility of notional relations), to which mathematics supplies precise and detailed answers. In this sense, the Ideas call the mathematical theories into existence (though we should not expect to discern the Ideas doing the calling until after the mathematical theory is complete.) Thus, the mathematical theories depend for their objectivity on the Ideas. The Ideas are objective in the sense that they are part of the deep structure of our engagement with the world, which means that they are part of the deep structure of us, or, what comes to the same thing, they are part of the deep structure of *our* world. Since our world is the only one we know, we may as well say that they are part of the deep structure of *the* world, so long as we remember that this insight is ontological, not ontic. It belongs to phenomenology, not to science.

4 Lautman's mathematical examples

Having sketched Lautman's view abstractly, I shall now consider his mathematical examples. The first of Lautman's two theses (*On the unity of the mathematical sciences*) takes as its starting point a distinction that Hermann Weyl made in his 1928 work on group theory and quantum mechanics. Weyl distinguished between 'classical' mathematics, which found its highest flowering in the theory of functions of complex variables, and the 'new' mathematics represented by (for example) the theory of groups and abstract algebras, set theory and topology (Lautman, 2006, p. 83–84). For Lautman, the 'classical' mathematics of Weyl's distinction is essentially analysis, that is, the mathematics that depends on some variable tending towards zero: convergent series, limits, continuity, differentiation and integration. It is the mathematics of arbitrarily small neighbourhoods, and it reached maturity in the nineteenth century. On the other hand, the 'new' mathematics of Weyl's distinction is 'global'; it studies the structures of 'wholes' (Lautman, 2006, p. 84). Algebraic topology, for example, considers the properties of an entire surface (how many holes?) rather than aggregations of neighbourhoods. Having quoted and illustrated Weyl's distinction, Lautman re-draws it:

> In contrast to the analysis of the continuous and the infinite, algebraic structures clearly have a finite and discontinuous aspect. Though the elements of a group, field or algebra (in the restricted sense of the word) may be infinite, the methods of modern algebra usually consist in dividing these elements into equivalence classes, the number of which is, in most applications, finite.[24]

---

[24]"[...] en opposition à l'analyse du continu et de l'infini, les structures algébriques

The chief part of Lautman's 'unity' thesis is taken up with four examples[25] in which theories of modern analysis (that is to say, analysis as practiced in the twentieth century) depend in their most intimate details on results and techniques drawn from the 'new', algebraic side of Weyl's distinction. In these four cases, algebra comes to the aid of analysis. Thus, Lautman transforms a broad historical distinction (between the local, analytic, continuous and infinitisic mathematics of the nineteenth century, and the new, 'global', synthetic, discrete and finitistic style) into a family of dialectical dyads (local/global, analytic/synthetic, continuous/discrete, infinitistic/finitistic). These pairs are not empty oppositions. They find their content in the details of mathematical theories that, though they belong to analysis, sometimes employ a characteristically algebraic point of view. In other words, the methods are algebraic but the results belong to analysis.[26] By this point, we have left the nineteenth century behind, and are concerned with analytic/algebraic 'mixtures' in contemporary (twentieth century) mathematics.

In his other major thesis, *Essay on the notions of structure and existence in mathematics*, Lautman gives his dialectical thought a more philosophical and polemical expression. Six chapters compose the body of this second thesis, the first three on 'structural schemas' (*schémas de structure*), the second three on 'origination schemas' (*schémas de genèse*). The three structural schemas are: local/global, intrinsic properties/induced properties and the (unfortunately titled) 'ascent to the absolute'.[27] The first two of these three schemas are pairs of the sort we saw in Lautman's 'unity' thesis. The 'ascent to the absolute' is a different sort of pattern; it involves a progress from mathematical objects that are in some sense 'imperfect', towards an object that is 'perfect' or 'absolute'. His two mathematical examples of this 'ascent' are: class field theory, which 'ascends' towards the absolute class field, and the covering surfaces of a given surface, which 'ascend' towards a simply-connected universal covering surface. In each case, there is a corre-

---

ont un aspect nettement fini et discontinu. Quelle que soit l'infinité des éléments qui constituent un groupe, un corps, une algèbre (au sens restreint du mot), les méthodes de l'algèbre moderne consistent le plus souvent à imposer à ces éléments une division en classes d'éléments équivalents, et à substituer ainsi à un ensemble infini la considération d'un nombre de classes qui, dans les applications, est le plus souvent fini" (Lautman, 2006, pp. 86–87).

[25] Dimensional decomposition in function theory; non-Euclidian metrics in analytic function theory; non commutative algebras in the equivalence of differential equations; and the use of finite, discontinuous algebraic structures to determine the existence of functions of a continuous variable (Lautman, 2006, p. 87).

[26] "[...] il est possible de retrouver dans les théories modernes de l'analyse les points de vue qui caractérisent l'algèbre[...] théories dont les méthodes sont algébriques mais les résultats s'étendent à l'analyse" (Lautman, 2006, p. 121).

[27] *La montée vers l'absolu.* The Hegelian resonance of 'the absolute' is a red herring; it seems to have deceived Bernays (1940, p. 20).

sponding sequence of nested subgroups (with the trivial subgroup mapping to the 'absolute' class field or surface), which induces a 'stepladder' structure on the 'ascent'. (Lautman introduces this idea with a brief discussion of the Galois correspondence, (Lautman, 2006, pp. 166–168).) This dialectical pattern is rather different to the others. The earlier examples were of pairs of notions (finite/infinite, local/global, *etc.*) and neither member of any pair was inferior to the other.[28] As we saw, Lautman argues that on some occasions, finite mathematics offers insight into infinite mathematics (think for example of the use of finite integer fields in the study of infinitely numerous natural numbers). In mathematics, the finite is not a somehow imperfect version of the infinite. Similarly, the 'local' mathematics of analysis may depend for its foundations on 'global' topology (as Lautman argues, Lautman, 2006, p. 85), but the former is not a botched or somehow inadequate version of the latter. Lautman introduces the section on the 'ascent to the absolute' by rehearsing Descartes's argument that his own imperfections lead him to recognise the existence of a perfect being (God). Man (for Descartes) is not the dialectical opposite of or alternative to God; rather, man is an imperfect image of his creator. In a similar movement of thought, according to Lautman, reflection on 'imperfect' class fields and covering surfaces leads mathematicians up to 'perfect', 'absolute' class fields and covering surfaces respectively. In short, the 'ascent to the absolute' introduces a different dialectical structure from the pairs of notions we saw hitherto. It has nothing in common with the patterns found in the three Platonic dialogues.

The three origination schemas are titled 'Essence and existence', 'Mixtures' and 'On the exceptional character of existence'. In the first two of these chapters, the structure of a mathematical domain gives rise to new mathematical objects; in the third chapter, Lautman considers cases where an object is shown to exist in virtue of exceptional properties that distinguish it from an established set of objects. Lautman does not address directly the general question of the metaphysical status of mathematical objects. He is, rather, interested in the way that mathematical structures and objects grow out of already existing mathematics. As he explains, the roles of 'originating structure' and 'created object' are relative; objects that owe their existence to the structure of another domain may themselves come to constitute the originating structure for some further class of entities (Lautman, 2006, p. 187). Part of his purpose is to oppose the view that there is nothing more to mathematical existence than the consistency of an axiom system. Lautman spends some time rehearsing the familiar technical difficulties entailed in attempts to prove the consistency of a system (Lautman, 2006, pp. 179–187). But his real claim is that mathematical entities do

---

[28]But cf. also (Barot, 2008, p. 12). Recall too that for Plato, the intellectual is superior to the sensual.

not depend for their existence on apparently arbitrary decisions to explore some sets of axioms but not others. Rather, mathematicians create new mathematical structures in the course of answering questions latent in the underlying extra-mathematical dialectical order. Here too, the dialectical patterns that he discerns are more richly varied than his references to Plato would suggest.

5 Blurring the ontological difference

Lautman concludes the *Essay on the notions of structure and existence in mathematics* with some remarks that, though they do not mention Heidegger by name, are clearly of a piece with the explicit discussion of Heidegger in *New research on the dialectical structure of mathematics* (Lautman, 2006, pp. 228–229). After a brief discussion of Plato (Lautman, 2006, pp. 230–234), Lautman ends his thesis with a statement of his credo, which he held to be true of mathematics and physics alike:

> The nature of reality, its structure and the conditions of its origination cannot be known except by returning to the Ideas that science embodies in its inner relations.[29]

As we saw, Lautman appeals to Heidegger in order to explain the relation between dialectics and mathematics. The whole point of *On the Essence of Ground* is to insist on the ontological difference, that is, on the distinction between the ontological and the ontic. The division of labour between the scientist and the philosopher depends on this distinction. The scientist uses ontic concepts to establish ontic truths; the philosopher reveals the corresponding ontology. Lautman insists on the distinction between dialectics and mathematics. If dialectic tries to find its own solutions to the problems it expresses, it will "mimic mathematics with such a collection of subtle distinctions and logical tricks that it will be mistaken for mathematics itself".[30] This, he suggests, is the fate of the logicism of Frege and Russell. Dialectical notions and ideas must find expression in mathematical examples. Expression in mathematical examples subjects an Idea to "a whole train of specifications, limitations and exceptions with which mathematical theories are constructed and confirmed".[31] So, for example, we might

---

[29]"La nature du réel, sa structure et les conditions de sa genèse ne sont connaissables qu'en remontant aux Idées dont la science incarne les liaisons" (Lautman, 2006, p. 234).

[30]"Une dialectique qui s'engagerait dans la détermination des solutions que ces problèmes logiques peuvent comporter, se verrait entraînée à constituer tout un ensemble de distinctions subtiles et d'artifices de raisonnement qui imiteraient a ce point les mathématiques, qu'elle se confondrait avec les mathématiques elles-mêmes" (Lautman, 2006, p. 228).

[31]"Il faut ensuite, pour que l'exemple supporte l'Idée, apporter à celle-ci tout un cortège de précisions, de limitations et d'exceptions où s'affirment et se construisent les théories mathématiques" (Lautman, 2006, p. 243).

look at the various mathematical concepts of completeness and closure, and recognise in them mathematical versions of the vague (and hence presumably dialectical) notion that a complex item might be self-sufficient or *sui generis*. To recall one of his detailed examples, Lautman invites us to see the mathematical relations between the intrinsic and relational properties of mathematical objects as a mathematical specification of the dialectic of same and other.

But now we have a problem. How are we to distinguish between the legitimate activity of seeking mathematical answers to dialectical questions, and the mistaken activity of making dialectics imitate mathematics? After all, historically, mathematics does not have fixed borders. For example, Euler thought that the Königsberg bridges problem lay outside mathematics, because "the solution is based on reason alone, and its discovery does not depend on any mathematical principle".[32] Formal logic lay outside mathematics for over two millennia (if we measure from Aristotle's *Prior Analytics* to Boole's 1847 *Mathematical Analysis of Logic*). Aside from the authority of Heidegger, Lautman's distinction between dialectics and mathematics depends on the "essential insufficiency"[33] of dialectical Ideas, that is, the fact that they cannot be understood except through the development of mathematical theory. However, this is also true of undeveloped or primitive mathematical concepts. The primitive concepts of continuity and infinity posed questions that were only properly answered through the contemplation of mathematical examples and the articulation of mathematical theories. For all that, continuity and infinity are clearly mathematical concepts, however primitive.

Lautman's own examples suggest that the line between dialectics and mathematics is neither clear nor stable. Look again at the diagram he takes from Stenzel (Figure 1). Lattices were not mathematical objects in Plato's day, but they are now. Are we to suppose that the underlying dialectical structure of Plato's arithmetic is itself an example of a mathematical concept (namely, lattice), which presumably has a dialectical basis of its own? Lautman explicitly rejects such regresses (Lautman, 2006, p. 232).

The second part of *New research on the dialectical structure of mathematics* is a pair of case studies that pick up the contrast between analysis and algebra that we first met in *Essay on the unity of the mathematical sciences*. In that early essay, we saw algebra coming to the aid of analysis. In these two cases, we see analysis (the mathematics of continuity) supplying proofs to number theory. The second case supports Lautman's argument for the unity of mathematics rather well: it is the use of the Rie-

---

[32]Letter 590 in Euler's *Opera Omnia*: quoted from Wilson (2008, p. 15). Euler wondered whether it might be what Leibniz meant by 'geometry of position'.

[33]"Insuffisance essentielle" (Lautman, 2006, p. 243).

mann zeta-function to investigate the density of primes. The first case is rather artificial: it is Hecke's proof of quadratic reciprocity. It is artificial because (as Lautman acknowledges) there is no need to call on analysis to prove this theorem. Of Hecke's proof of quadratic reciprocity, Lautman writes:

> The analytic tool, that is to say, functions, serves to demonstrate an arithmetical result because the structure of the tool and that of the result both participate in the same dialectical structure, which poses the problem of the reciprocity of roles between mutually inverse elements.[34]

This presents two problems for the distinction between dialectics and mathematics. First, reciprocity is a kind of symmetry. The symmetry that obtains between these mutually inverse elements hardly requires the apparatus of group theory, but it is, nevertheless, a mathematical concept, as indeed are the relata. Second, the proper relationship between dialectics and mathematics appears to have been reversed. In a letter to the mathematician Maurice Fréchet, Lautman explained:

> It is insofar as a mathematical theory supplies an answer to a dialectical problem that is definable but not resolvable independently of mathematics that the theory seems to me to participate, in the Platonic sense, in the Idea with regard to which it stands as an Answer to a Question.[35]

In principle, then, dialectics stands to mathematics as question to answer, but here a mathematical question (why does this body of analytic theory serve to prove that arithmetical result?) gets a dialectical answer. In any case, 'participates in the same dialectical structure as' is a symmetric relation, but the tool-result relation is not. He gives other examples in *New research on the dialectical structure of mathematics* in which (he claims) "the convergence of different mathematical theories results from the affinity of their dialectical structures",[36] but he elsewhere gives examples of mathematical theories that share dialectical structures (such as all the same/other examples) but do not show any sign of convergence.

---

[34]"l'outil analytique, c'est-à-dire les fonctions, sert à démontrer un résultat arithmétique, parce que la structure de l'outil et celle du résultat participent l'une et l'autre d'une même structure dialectique, celle que pose le problème de la réciprocité de rôles entre éléments inverses l'un de l'autre" (Lautman, 2006, p. 248).

[35]"C'est dans la mesure où une théorie mathématique apporte une réponse à un problème dialectique définissable mais non résoluble indépendamment des mathématiques que la théorie me paraît participer, au sens de Platon, à l'Idée vis-à-vis de laquelle elle est dans la même situation que la Réponse par rapport à la Question" (Lautman, 2006, p. 260).

[36]"la convergence des théories mathématiques différentes résulte de leur affinité de structure dialectique" (Lautman, 2006, p. 250).

In short, the claim that science and phenomenology treat of different concepts collapses in mathematical practice.[37] Symmetry (for example) is a mathematical concept, but it can also function as a dialectical notion in Lautman's sense—it is one of the notions that shape our questions. If it does so explicitly, it may also function as a heuristic in the sense of Polya. Lautman does not mention heuristics; rather, he insists that we should not expect to discern a dialectical question in advance of arriving at its mathematical answer. However, this overlooks the fact (which Lautman elsewhere insists on) that the same dialectical notions and Ideas may feature in different mathematical theories. As a notion or Idea recurs in various mathematical theories, it may become an explicit part of the mathematical culture and thus begin to function heuristically. Recognising that a concept can serve on either side of the dialectical/mathematical distinction would be consistent with what we find in the later Plato. The 'same/other' dyad may have a deep ontological role, lending intelligibility to concepts as diverse as 'place' and 'intrinsic/extrinsic', but it also has a function in unremarkable empirical questions like "Is that the same dog as I saw yesterday?". Consequently, if we wish Lautman to enrich our own philosophy, the first move should be to give up the 'ontological distinction'. This thought can even find some support in Heidegger. In a 1936 lecture *Modern Science, Metaphysics and Mathematics*, he writes:

> The greatness and superiority of natural science during the sixteenth and seventeenth centuries rests in the fact that all the scientists were philosophers. They understood that there are no mere facts, but that a fact is only what it is in the light of the fundamental conception [...] the present leaders of atomic physics, Niels Bohr and Heisenberg, think in a thoroughly philosophical way [...](Heidegger, 1962, p. 272).

It is hard to see what advantage philosophy could bring to science if the ontological distinction stands between them. Heidegger's claim that great scientists are also philosophers suggests that they do not respect the 'ontological distinction' in their practice.

This modification would also give Lautman a reply to a criticism from one of his closest colleagues. At a meeting in February 1939, Lautman insisted that the objectivity of mathematical theories depends on their participation in non-mathematical Ideas that dominate them. Also present was Jean Cavaillès, who remarked, "Personally, I recoil from positing something else which would dominate the actual thought of mathematicians, I see necessity in the problems [...]"[38]. Giving up the 'ontological distinction' would

---

[37] For an independent argument with a similar conclusion, cf. Barot (2008, p. 14–17)

[38] "Personnellement je répugne à poser une autre chose qui dominerait la pensée effective du mathématicien, je vois l'exigence dans les problèmes[...]" (Lautman, 2006, p. 263).

allow Lautman to reply that dialectical Ideas do indeed dominate mathematical theories, but the Ideas, the theories and the domination are all part of mathematical thinking. Thus, talk of 'domination' notwithstanding, no extraneous constraint cramps the thought of mathematicians.

Bibliography

Barot, E. (2008). L'Objectivité Mathématique selon A. Lautman: entre Idées dialectiques et réalité physique. *Cahiers François Viète*, 6:3–28.

Bernays, P. (1940). Review of *Essai sur les notions de structure et d'existence en mathématiques* and *Essai sur l'unité des sciences mathématiques dans leur développement actuel* by Albert Lautman. *Journal of Symbolic Logic*, 5(1):20–22.

Cornford, F. M. (1924). Review of *Zahl und Gestalt bei Platon und Aristoteles* by Julius Stenzel. *Classical Review*, 38(7–8):209.

Heidegger, M. (1925). Platon: Sophistes. Lecture course Wintersemester 1924/25; published as Schüßler (1992). Page numbers refer to the English translation (Heidegger, 2003).

Heidegger, M. (1962). *Die Frage nach dem Ding. Zu Kants Lehre von den transzendentalen Grundsätzen*. Max Niemeyer, Tübingen. Page numbers refer to the English excerpt in Heidegger (1993).

Heidegger, M. (1967). *Wegmarken*. Vittorio Klostermann, Frankfurt a. M. Page numbers refer to the English translation (Heidegger, 1998).

Heidegger, M. (1993). *Basic writings*. Routledge, London.

Heidegger, M. (1998). *Pathmarks*. Cambridge University Press, Cambridge.

Heidegger, M. (2003). *Plato's Sophist*. Indiana University Press, Bloomington, IN.

Lautman, A. (2006). *Les mathématiques, les idées et le réel physique*. Vrin, Paris. Introduction and biography by Jacques Lautman; introductory essay by Fernando Zalamea; preface to the 1977 edition by Jean Dieudonné.

Schüßler, I., editor (1992). *Martin Heidegger. Gesamtausgabe. II. Abteilung. Vorlesungen 1919–1944. Band 19. Platon: Sophistes*. Vittorio Klostermann, Frankfurt a. M.

Shorey, P. (1924). Review of *Zahl und Gestalt bei Platon und Aristoteles* by Julius Stenzel. *Classical Philology*, 19(4):381–383.

Trevaskis, J. R. (1967). Division and its relation to dialectic and ontology in Plato. *Phronesis*, 12(2):118–129.

Wilson, R. (2008). Euler's combinatorial mathematics. *Bulletin of the British Society for the History of Mathematics*, 23(1):13–23.

François, K., Löwe, B., Müller, T., Van Kerkhove, B., editors,
Foundations of the Formal Sciences VII
Bringing together Philosophy and Sociology of Science

On the philosophical talk of scientists

Hauke Riesch[*]

Judge Business School, University of Cambridge, Trumpington Street, Cambridge, CB2 1AG, United Kingdom
E-mail: hr277@cam.ac.uk

1 Introduction

This paper analyses how scientists talk and write about philosophical topics, as part of a larger project on scientists' views of philosophy of science. It aims to describe how scientists themselves think and learn about the nature of science, and what they would like other people to learn about it. A total of 30 popular science books were analysed for how they treat philosophical topics on the nature of science. Additionally, 40 academic scientists were asked in a series of semi-structured interviews questions based on the philosophical topics that were found discussed most often in the books. In the interviews, five philosophical topics were dealt with in detail: The demarcation question of "what is science", the philosophies of Popper and Kuhn, Occam's razor and reductionism, which reflect the most common philosophical themes in the popular science books. This paper will focus on two of these topics as characteristic of scientists' talk of philosophy, the general question of what distinguishes science from other endeavours, and the philosophy of Karl Popper.

In interpreting the books and the scientists' responses on these topics, I will introduce social identity theory to argue that philosophy can be used to rhetorically draw social boundaries and to define social identities around science. This talk surrounding the various philosophical categories however often hides a big variation in actual philosophical opinion, which is set slightly apart from how the philosophy itself is discussed.

---
[*]Many thanks to Hasok Chang, Brian Balmer, Berangere Bacquey and the scientists who participated in the study. The research was funded by the Economic and Social Research Council.

Received by the editors: 20 January 2009; 12 October 2009.
Accepted for publication: 5 March 2010.

Next to being a contribution towards a fuller sociological understanding of how science thinks of itself, this study is also intended as an argument for using the sociological study of the philosophy of scientists for the help of the philosophy of science itself, and for making it relevant to how science is understood—by the public as well as ultimately scientists themselves. It shows how the two disciplines of sociology and philosophy of science can complement each other by using a science and technology studies approach to the philosophy of science by scientists.

I will adopt the convention from science education studies and use "the nature of science" as an umbrella term to cover historical, philosophical and sociological studies of science. The term can be slightly misleading because, as these disciplines have shown, there is no universally accepted definition of the nature of science, and as I will show in this paper, neither is there a universally accepted definition within the scientific community. Thus, when I asked scientists about how they see the nature of science, I was after their own personal perspective about science. I use the term "scientific method" to mean the branch of the philosophy of science that concerns itself with analysing how science is and should be done. I will therefore use the term in a way that recognises that there are many different ideas of what precisely scientific method consists of.

Studies of scientists' philosophies have been done before, although with various aims in mind, using different methodologies and theoretical frameworks, and looking at different philosophical topics in detail. All of this, plus the fact that these studies were made from the perspectives of different disciplines, makes pooling their results and conclusions very difficult.

1.1 Public understanding of science and science education

Research in the related disciplines of science education and "public understanding of science" (PUS), has frequently argued for the virtues of teaching philosophy (along with history and social aspects of) science. In the tradition of public understanding of science survey research on scientific literacy, e.g., researchers such as Jon D. Miller argue that an understanding of the nature of science is essential for basic scientific literacy, and the question of "what is science" is usually included in public science literacy surveys (e.g., Miller, 1987). Work in this tradition has come under criticism within PUS as conceptualising the public as simply deficient in scientific knowledge so that the science communicator's job is merely to provide the facts (notably called by Brian Wynne the "deficit model of PUS", cf. Wynne, 1992 for his critique of the deficit model), which neglects the facts that public knowledge of science is contextualised within society, and sometimes even more relevant than the experts' judgements. The traditional "deficit model" preoccupation with the literacy on the nature of science has come under similar criticism, e.g., Bauer and Schoon (1993) argue that Miller's assessment of literacy in scientific method is too heavily biased towards how Miller himself understands it (they charge it for being too Popperian). Miller replied that his idea of scientific method is just that of ordinary scientists. Since his PUS surveys are designed "to measure the level of public understanding of the scientific approach as understood and used by scientists", Bauer and Schoon are missing the point: "The question was not created to gauge the public's views on the philosophy of science" (Miller, 1993, p. 237). This answer of course somewhat begs the question of what scientists really do think about scientific method. Neither Miller nor Bauer and Schoon had an answer to that, and through its shift of focus to contextual models of understanding science, PUS has now largely abandoned the idea of scientific literacy and with it preoccupations of what philosophical conceptions of science should be taught.

That question however has stayed relevant for science education, which has a large body of literature devoted to the practical and theoretical aspects of teaching philosophy and history of science in science education (among many others there are Matthews 1994; 1998; Abd-El-Khalick and Lederman, 2000; Lederman, 2007; Koponen, 2006). Within this tradition of research, there exist also a lot of empirical studies on various groups' conceptions of the nature of science, mostly concentrating on science students and their teachers/educators (Lederman, 1992 for a review of the literature on teachers' conceptions, Driver et al., 1996; Dogan and Abd-El-Khalick, 2008). Less prominently there are also studies on philosophers themselves (Alters, 1997) and lastly scientists and other experts (Osborne et al., 2003; Wong and Hodson, 2009).

1.2 Philosophy and history of science

Scientists' philosophical ideas have also been studied by philosophers, both through the thought that their shop-floor insights may help the development of philosophy itself, and because some philosophers think it is reasonable that they should not stray too much in spirit from how scientists think about their subject. Thus Bailer-Jones (2003) has interviewed scientists on their ideas of scientific models, while a research group at Pittsburgh has recently sought to inform philosophical thinking about the gene by asking scientists themselves about how they understand the philosophical issues around the concept (Stotz et al., 2004). Much more informal that these attempts is a philosopher's attempt through his regular popular philosophy column to canvass readers of Physics World on their assumptions concerning realism (Crease, 2001, 2002).

This rather sparse research on contemporary every day working scientists stands in contrast to numerous studies by philosophers that focus on historical case studies (which however usually look at scientists' actions rather than their thoughts) that try to test philosophical theories on science by looking at historical developments. The literature is huge, but well represented by research program of Donovan et al. (1992), or the analysis on how scientists evaluate predictions in Brush (1989).

1.3 Sociology of Science

Lastly, scientists' opinions on philosophical topics has been an occasional research subject of sociologists of science. One study that stands out is from Gilbert and Mulkay's (1984) hugely influential book analysing interviews with a group of biochemists. The scientists' talk of Karl Popper was written up as a separate article for a philosophy audience that did not make it into the famous book (Mulkay and Gilbert, 1981). They identified what they call an empiricist and a contingent interpretative repertoire in the way scientists talk about science: the empiricist repertoire is a collection of interpretative phrases and devices scientists use when talking more formally about science, and in which science is represented as logical, rational and following empirical and logical philosophies such as that of Karl Popper. The contingent repertoire by contrast was used more often when scientists were talking informally about science, where it was acknowledged that science has a human and contingent side to it. In the case of their talk about Popper, scientists would, e.g., state that science follows the methodology of falsificationism. At the same time when talking about particular scientists such as their rivals, they would talk about human fallibilities, e.g., that this scientist talked about being a Popperian, while he clearly failed to put that into action. Working from a similar theoretical perspective, Potter (1984) has analysed psychologists' discourse of Thomas Kuhn at a conference he attended. Some other sociological work has analysed scientists' written discourse on philosophy; Sovacool (2005) surveyed astronomy papers for their references to Popper, while Nieman (2000) and Turney (2001) analysed philosophical references in popular science books.

In this paper I will present a study that complements the ones surveyed above, and hopefully add a perspective that combines these three concerns. Scientists learn foundational and philosophical issues about science primarily from other scientists, as became very clear in my interviews. How philosophy gets represented by scientists, what uses it is being put towards, and how that influences scientific thought is interesting from a sociological and philosophical point of view, because it shapes science at the same time as it shapes lay philosophical thinking, and the way scientists believe science should be taught.

1.4 Methods

I conducted four pilot interviews based on a convenience sample of physicists and astronomers, in which I tested some prior ideas of how scientists talk about philosophical topics and to find which topics they found interesting and/or controversial. The pilot interviews guided the topics I selected for the main interview study, and directed me to which topics to pay most attention to in the books. The pilot interviews were later coded and analysed together with the main interviews.

After the pilot, I have closely read 30 recently published and acclaimed popular science books and analysed how they represent themes from the philosophy of science in general. Next to building up a picture of how the authors think science differentiates itself from other endeavours, I have also tracked other philosophical themes that have come up in the books most; these were the philosophers Kuhn and Popper as well as the concepts of reductionism and Occam's razor or simplicity as a value in science. I have decided to look at popular science because the genre gives a unique opportunity to scientists for explaining why and how their science was conducted, away from the strict constraints of the technical and textbook literature. Furthermore, because popular science is specifically aimed at explaining science to intelligent non-scientists, it contains a lot of explanations of how science in general works, and this inevitably includes many philosophical observations. Popular science is therefore a rather unique genre for giving us an insight into how the authors view the philosophy of science (as argued by Turney, 2001).

Relevant pages from the books and the transcripts from the interviews were analysed using qualitative data analysis software, Nvivo and Atlas.ti. In order to build up a corpus of books that were unquestionably scientific, I chose from the shortlists of the Aventis prize for popular science (called the Rhone-Poulenc Prize until 1999 and since 2006 the Royal Society prize). To get a selection of contemporary books I selected books from between 1998 and 2004; books not written by scientists and books about mathematics were filtered out (cf. the appendix for a full list of books that I have included in the corpus). The precise definition of who is a scientist for the purposes of constructing the corpus is by necessity slightly vague. I have striven to be as inclusive as possible and include authors who have some experience of conducting science themselves, and have therefore opted to select authors who either have a doctorate in a science discipline (again, trying to be broad with me definitions and thus including psychology as well), or those who have published research in peer reviewed science journal.

Then, 36 scientists were interviewed on the topics that came up in the books, as well as a few more general question about how they themselves have learned about what is science and how it should be done. The scientists were all chosen from university departments of physics, chemistry and biology, about two thirds based in the UK, and a third in Paris. In the sections below, interviewees will be marked by a unique number, their gender and seniority and the country they are based in (a few interviewees had experience of working in both Anglo-phone and French environments, in which case I have marked this as well). The pilot interviews were included in the final analysis and marked as numbers 1 to 4. French scientists were given the choice to hold the interview in French, in which case I will present my own translation in the main text below and provide the original as a footnote.

In the interviews I have asked the scientists to give me their idea of how science works, and what distinguishes science from other pursuits. When a scientist answered merely that it consists of following the scientific method, I have asked them to explain what exactly the scientific method is. I have also asked how they acquired their current opinions (e.g., through tuition, reading philosophy or introspection). Afterwards I asked targeted questions on Occam's razor, reductionism and the philosophies of Kuhn and Popper.

By comparing scientists' views in interviews with how science presents itself publicly in popular works, the study first presents us with a comparison that can ask whether science's publicly visible face is giving us a portrait of science that is consistent with the way ordinary working scientists think about it, though I will not concentrate on that effect on this paper. Second, however, the comparison can highlight for us the way the intended audience can influence the way science is portrayed. Because popular science is highly visible not only to the public but to other scientists, it affords the author an opportunity to be seen the way he or she wants to be seen, and therefore rhetorical considerations about the effect adherence to a particular philosophy can have become important.

Through the decision to perform a qualitative study, this project is not intended to show a representative sample of scientists' opinions (although I have tried to construct my corpus as representatively as possible), but rather to give a perspective of the different opinions that exist within science regarding the philosophy of science. One of the ideas at the beginning was to supplement the study with a quantitative survey, but it soon became apparent that especially abstract topics like those of philosophy can be interpreted so widely that simply counting the number of "Popperians" will not give a worthwhile insight into the thinking of working scientists. The pitfalls of such a survey are well demonstrated by Crease's (2001 and 2002) admittedly light hearted and informal sampling of physicists' opinions on realism which gave us the phenomenal result that 7% of physicists believed atoms were not real, and 3% even thought the earth did not exist. This does not tell us the scientists' interpretations of science. Are these 3% of scientists idealists? We simply don't know.

The analogy that Bauer et al. (2000) give to characterise the difference between qualitative and quantitative research is:

> If one wants to know the colour distribution in a field of flowers, one first needs to establish the set of colours that are in the field; then one can start counting the flowers of a particular colour. (Bauer et al., 2000, p. 8)

This paper is in an attempt to find out the different categories in which scientists talk and write about the philosophy of science, and offer some tentative interpretations of what is going on. Generally when referring to the numbers of books, or interviewees who responded a certain way I will use qualitative terminology such as "some", "a few" or "most". Giving precise numbers of the number of occurrences of particular codes even in a fairly large qualitative corpus is appropriate only in limited circumstances. Very often, e.g., when it came to assessing how many scientists agreed with Popper, there are many borderline and indeterminate cases so that giving precise numbers can be misleading (cf. also Hammersley, 1996, p. 161 ff. for a discussion on the use of quantitative terms in qualitative research). Many comments also came up within the free-flowing conversational format of the interview, and some conversations covered ground not covered in exactly the same way by others. Therefore the numbers that I will provide should be interpreted with some caution.

2 How philosophy is used

2.1 Philosophical asides: philosophy as authority and boundary marker

Almost every single popular science book had at least one comment, phrase or even paragraph on what is science, or how it works or should work. Popular science books are often an account of what scientists have found out. It is natural to accompany these accounts by outlining how things were found out to work like that, and why that way of finding things out was so persuasive. One kind of comment is that of a more general philosophical discussion on what counts as good evidence, when to abandon or adopt a theory or hypothesis, or what sets science apart from other endeavours where knowledge is not so secure, such as (according to the tastes of the author) sociology, psychology or theology, and also what sets it apart from fraudulent, pseudo or fantastic science. These accounts are usually longer (in some cases up to chapter or even book length), and often feature specific pet philosophers (usually Popper and Kuhn) or philosophies (such as falsificationism, reductionism or Occam's razor.

The most pervasive category of philosophical comments however is the smaller aside that can appear in almost any context in popular science books. They are generally very short and in a way merely shift the appeal to authority that any presentation of facts rely on in popular science, to an appeal to authority that "this is just how things work in science". Instead of explaining how science works, they tell us how science works. Nieman (2000) analyses similar comments in his characterisation of the uses of philosophical remarks in popular physics. Nieman discusses what he calls "pithy" philosophical definitions that scientists give in their popular accounts as a kind of rhetorical boundary work (Gieryn, 1995), and that in popular science "discourse on the meaning of science is more concerned with the defending or capturing of territory than exploring the metaphysical subtleties of knowledge about nature" (Nieman, 2000, pp. 167–168).

Pithy comments, where something is shown to be good science because it conforms to the norms of science, occurred very often, whatever the author thinks that is: it is falsifiable, verifiable, it predicts things, is based on meticulous or rigorous testing, observing, peer reviewing. The mechanisms of that method are not discussed further, it suffices to show that the science in question conformed (or the pseudo-science failed to conform) to what the author holds as good scientific practice, e.g., when an author wanted to show "new-age dreaming" as not being science because it fails Occam's razor:

> The best theories are rendered lean by Occam's razor, first expressed in the 1320s by William of Occam. He said, "What can be done with fewer assumptions is done in vain with more." Parsimony is a criterion of good theory. With lean, tested theory we no longer need Phoebus in a chariot to guide the sun across the sky, or dryads to populate the boreal forests. The practice grants less license for New Age dreaming, I admit, but it gets the world straight. (Wilson, 1998, p. 56)

In an even pithier aside to scientific method, which is much more typical because it makes only a vague allusion to the philosophical thinking behind the science, Sapolsky discusses a sociobiological hypothesis about the value of kidnapping in baboon society.

> The alpha male is about to pound you [i.e., a threatened baboon]. You don't grab just any kid, you grab someone who he thinks is his kid. Mess with me and your kid gets it. Kidnapping, hostage taking. Pretty clever. The idea generated all sorts of predictions. (Sapolsky, 2001, p. 100)

The crucial part here is that the idea generated predictions, which is one of the more frequently voiced attributes of a good hypothesis, and Sapolsky goes on to argue that these predictions were then put towards the evidence to see if they supported the sociobiological hypothesis:

> The sociobiological model has been supported only to some extent
> by the data. Appendices have been added on to the theory. ...The
> debate rages on, keeping primatologists off the dole. (Sapolsky, 2001,
> p. 100)

Sapolsky here reveals a vaguely hypothetico-deductive stance, where it is important for a hypothesis to predict things which will then in turn be checked by experiment or observation. Sapolsky finishes this section by admitting that the available data has not yet been able to settle the dispute, with a humorous but resigned verbal shrug.

Although it is possible to characterise to an extent what the prevailing philosophical opinions were among the books, the context in which most philosophical remarks were made were often as a rhetorical device for quickly demonstrating why something being discussed is not proper science, without the need to go into particular details. These comments do not necessarily add up to a coherent or complete view on how science works. Nor do they necessarily represent what the author thinks is the most important distinguishing feature of science: The actual philosophical point may have been chosen to show that bit of science being discussed in a most favourable light (or the other way around for pseudo-science), and the highlighted bit of philosophy may therefore have been chosen for convenience.

There were no real equivalent to these philosophical asides in the interviews, because I asked the question of what is science directly. However, many scientists had a ready answer to the question, that they gave without hesitation but which was considerably qualified during the later discussion, suggesting that here philosophy again fulfilled a rhetorical function as an appeal to authority which was taken as given, instead of being questioned. This was particularly the case for the philosophy of Karl Popper, which will be analysed in more detail in the following section.

2.2 Philosophy as an identity: The case of Popper

In social identity theory (Tajfel, 1981; Tajfel and Turner, 1986; Hogg and Abrams, 1988), which has by now become a standard tool to make sense of group interactions within social psychology (cf. Brown, 2000 for a critical review) , individuals within a group enhance their own self-esteem by self-categorising themselves as conforming to group norms and values, which are usually perceived as positive and desirable attributes. In the process of categorising ourselves we tend to accentuate those aspects of our own attributes and values which conform with those of our group (the ingroup), while we simultaneously downplay those that are in conflict with it. At the same time, when considering the attributes of people outside of the group (the outgroup), we tend to accentuate the negative aspects and downplay the positive ones and those they actually have in common with the ingroup.

Social identity theory has explicitly been formulated primarily to explain stereotyping and discriminative behaviour (as explained in a biographical note by Tajfel, 1981), but also accounts for the more positive aspects of ingroup identification, and the discursive and rhetorical practice with which identification is negotiated. In experimental settings social identity theory has a amassed a considerable amount of evidence in its favour though it still lacks in applications to real life case studies (as argued by Huddy, 2001).

Probably unsurprisingly, the one philosopher that was mentioned most often in both books and spontaneously in the interviews, was Karl Popper. Popper was mentioned spontaneously in answer to my first question on what distinguishes science by seven interviewees, and only seven interviewees said that they have never heard of him. Scientists' allegiances to Popper were however curious: the reason that I introduced social identity above, is that Popper seems to have become a group norm for science, so that following Popper has become a value that many scientists think they fulfil by virtue of belonging to that group. Of the eight books that mentioned Popper, only one (Mayr, 1997) was negative towards him, though he was negative towards all philosophers. Through the accentuation effect, and the fact that as an abstract philosophical set of statements Popper's philosophy is already open to a fairly wide interpretation, it has become possible for a scientist to identify with Popper's philosophy, almost regardless of his/her actual philosophical opinion.

This identification effect was particularly visible in the books. Popper (or falsificationism) would, e.g., be mentioned in a pithy, authoritative way as outlined above, where the mere mention of a scientific episode following Popper was seen to be enough to show that something is scientific. The natural historian Richard Fortey, e.g., backs up his discussion on ad hoc explanations by remarking that they are

> anathema to all those brought up with the scientific and philosophical rigour of Karl Popper and Ernst [sic] Nagel. Scientists do not trot out ad hocs the way a magician pulls flowers out of a top hat; it is not considered proper behaviour. (Fortey, 2000, p. 241)

Note also how Fortey includes the for philosophers slightly contrary stances of Popper and the post-positivist philosopher Nagel as both exemplifying good scientific thinking.

The conflation of Popper and the logical positivists went even further with Stephen Hawking:

> Any sound scientific theory, whether of time or of any other concept, should in my opinion be based on the most workable philosophy of science: the positivist approach put forward by Karl Popper and others. (Hawking, 2001, p. 31)

Here Hawking does not actually interpret positivist philosophy by widening it to include Popper's falsificationism, instead he seems to describe a fairly consistent logical positivist outlook, and then thinks Popper shared it:

> If one takes the positivist position, as I do, one cannot say what time actually is. All one can do is describe what has been found to be a very good mathematical model for time and say what predictions it makes. (Hawking, 2001, p. 31)

Even authors who were very knowledgeable about Popper's philosophy, and who wrote about him at length, such as David Deutsch (1997) showed that it was possible to identify with Popper while at the same time criticising his philosophy. After all, there is no rule about how much we have to actually agree with Popper in order to be Popperians. In one chapter he sets up a fictional debate between a "crypto-inductivist" and himself about the validity of Popper's philosophy, and inductivism in general, which deserves a slightly longer quotation:

> CRYPTO-INDUCTIVIST: ... You make a careful distinction between theories being justified by observations (as inductivists think) and being justified by argument. But Popper made no such distinction. And in regard to the problem of induction, he actually said that although future predictions of a theory cannot be justified, we should act as though they were!
>
> DAVID: I don't think he said that, exactly. If he did, he didn't really mean it.
>
> CRYPTO-INDUCTIVIST: What?
>
> DAVID: Or if he did mean it, he was mistaken. Why are you so upset? It is perfectly possible for a person to discover a new theory (in this case Popperian epistemology) but nevertheless to continue to hold beliefs that contradict it. The more profound the theory is, the more likely this is to happen.
>
> CRYPTO-INDUCTIVIST: Are you claiming to understand Popper's theory better than he did himself?
>
> DAVID: I neither know nor care. The reverence that philosophers show for the historical sources of ideas is very perverse, you know. In science we do not consider the discoverer of a theory to have any special insight into it. On the contrary, we hardly ever consult original sources. They invariably become obsolete. (Deutsch, 1997, pp. 156–157, original emphasis)

In the interviews, however, while Popper was indeed very well recognised, that alone did not mean that his philosophy found much favour.

Some scientists had specifically built their whole understanding of science around Popper, though that was a fairly rare reaction. One scientist actually decided that Popper had in fact described science very well, after having heard negative opinions about Popper in his own undergraduate philosophy education:

> [In the philosophy course] Popper was a dirty word [laughs]. Well, that's how I remember it, I mean [...] I think the general idea was that science is more complicated than that. And that a lot of it is social influences and so on and so forth. But now I've been practising science quite a lot longer than I had then, [...] you know, clearly falsification is absolutely key, if [inaudible] going to have to test hypotheses, then falsificationism is what a lot of science is about. (12 Senior, Biology, Male, UK)

Very negative reactions towards Popper however also occurred. During a discussion of science in general during one of the pilot interviews, this scientist showed a lot of dissatisfaction with Popper:

> I think Popper's rejection of induction as a means of developing scientific theories and models is just crass in the extreme. (1 Senior, Physics, Male, UK)

Most reactions to Popper however were positive as well as negative. While Popper himself was assessed positively, most of those who had heard of Popper and had mulled over the philosophy of falsificationism applied to their own day to day scientific life, decided that there has to be more to science than just falsificationism, and that things like verification and induction have to have a role as well in any philosophy (20 interviewees in total).

But there can be a lot of similarity between agreeing with a philosophy and therefore in a way identifying with it, and on the other hand rejecting it while holding a fairly similar opinion. This is often manifested by followers of Popper still ascribing a large role to verification as well as falsification in science.

> Scientific method is that within some system you make hypotheses, you make testable hypotheses, you test them, and you keep the ones that are verified, and you throw out the ones that are falsified. (32 Senior, Physics, Male, France/En)

This scientist was very aware of Popper, admitting though that he does not know much of his work. Asked whether he accepted falsificationism as a philosophy, he replied, laughing: "Nothing else to say. I buy that one!". This shows that even when Popper has clearly influenced a scientist's thinking towards science, his precise teachings are not necessarily taken over wholesale. In this case it is, as the scientist admits himself, because he does not know that much about Popper's philosophy, and has only taken on board those aspects of it that he heard being discussed informally and those he agrees with.

The role he ascribes simultaneously to verification and Popper's philosophical authority is rather fascinating. Having explained a particular case from his own experience where he thinks to have verified something, he goes on to imagine Popper's response:

> So I think for me, from a more or less a logic layperson's point of view, we verified the hypothesis [...]. I guess Karl Popper is going to tell me that it could be many other things because we haven't tested every possibility [laughs]. (32 Senior, Physics, Male, France/En)

While the logic of science according to Popper may dictate rejecting verification, actual science manages quite well with it. Through this intervention of the imaginary Popper, this scientist manages both to defend the way he does his science, and yet defer to the philosophical authority of Popper, to which he interestingly, if mildly sarcastically, subjects his own scientific work. Regarding philosophy of science and logic, he regards himself a layperson.

I have argued above that some of the reason that popular science authors like Deutsch are prepared to identify themselves with Popper's philosophy while still disagreeing with it to some extent, may signify the adoption of Popper as an identity marker which shows the authors conform to what they think is proper scientific attitude, i.e., supporting Popper. Scientists who believe that Popper embodies rational and scientific thought, such as Fortey, will believe that they follow Popper's philosophy as long as they also believe that they are doing their science properly (as most scientists would obviously do). Discrepancies between the philosophy of Popper and the scientists' actually held philosophy can very easily be explained, either by the scientist having only an incomplete understanding of what Popper stood for as in the case of Hawking for whom Popper becomes a logical positivist, or by arguing that Popper would probably have agreed, as in the case of Deutsch.

In the interviews this identification is evident as well, though somewhat weaker, with respondents who reacted very positively towards Popper, but then also disagreed with his philosophy, or at least argued that there must be more to science than just falsification. Some of them also argued, similar to Deutsch, that surely Popper would have agreed with them, or that he did not really mean it like that. Throughout the interviews the interviewees also often qualified their remarks by saying that they were not well qualified to speak on these philosophical matters, which in some sense also gives the authority to speak of such matters back to the philosopher.

By these mechanisms, the generally (though by no means universally) positive attitude towards Popper hides a great variation in the way scientists thought about science, and the next section will try to chart and categorise these views.

3 Scientists' philosophical opinions on science

As I have argued, identifications with particular philosophers or philosophies by scientists may not necessarily translate into accepting them wholesale, so any survey of scientists' actual lay philosophies must try to look deeper than that. In the books that was naturally very difficult, because I couldn't ask direct questions. However, the interviewees have, despite the very limited amount of official philosophies they held to, shown a very wide range of opinions of how science works, and while this alone will not be surprising, I have found that these opinions map out differently than discussed in most philosophy textbooks, and their popular science representations.

The three overarching themes are familiar: Science is hypothetico-deductive, science is inductive, and science is a social enterprise. These however are discussed on slightly different terms than is usual in philosophy.

3.1 Science is hypothetico-deductive

Scientists most often, both in the interviews and as far as can be seen from the books, held that science works by proposing hypotheses, checking what their consequences are, and then proceeding to test them. This was the most frequent idea about science in the interviews. It also of course, maps onto the discussion on Popper above. Just as was the case with the books, there were several ways the respondents voiced the idea. Followers of Popper, e.g., unsurprisingly, emphasised the falsification aspect of the testing (ten in total), while many others spoke of verifying or even of both verifying and falsifying (ten), or that science is distinguished by being testable in some unspecified form (twelve). Others (ten, again) emphasised the idea that science predicts phenomena, without necessarily mentioning that those predictions will be put to test, though that was usually implied.

However, within a sea of comments such as that what distinguishes science is that it is testable, there were also some expressions of disquiet. First of all, even the idea of theories, and research that is driven by hypotheses was questioned:

> Respondent: I'm not entirely happy with theories, that, the business of hypothesis driven research, I find a little bit uncomfortable sometimes, because there are other ways of doing good research.
>
> Me: Such as?

> Respondent: I like the, the "this is an interesting question, let's [inaudible]" approach. "I wonder what would happen if ..." (8 Senior, Biology, Female, UK)

The complaint that there is more to science than just theory testing, was made very often (also often in the discussions following falsifiability discussed above). However, though this reservation was often voiced during the discussion on Popper, some scientists, like the one quoted above, have even brought this point forward within the initial discussion of what science is, without any prompting of mine. This conscious counter-positioning of course also signals that this scientist knows that the dominant scientific discourse on method follows hypothetico-deductivism.

3.2 Science is inductive

There were also comments that emphasised the accumulation of facts in science (seven interviewees said that science is a collection of facts or generalisations; twelve emphasised observation and/or experimentation). People who made these remarks always pointed out that science should of course not only consist of collecting facts or merely observing, but that a lot of characterisations of science miss out on this rather fundamental aspect. Therefore, for some scientists, induction had an important role to play in science. For example one biologist argued that an important part of science was laborious fact-finding which has to be done before hypotheses can be constructed at all.

> I think this aspect of science also this kind of accumulative ... accumulation of limited, but somehow useful, knowledge is an important part of science. (36 Senior, Biology, Male, France)

Very close to the idea that science is about collecting facts, is the inductive argument that science arrives at conclusions by generalising from the facts it observes:

> That, too is a thing common to all scientists, that from a particular thing we try to get a generality out, I think. (29 Early Career, Physics, Male, France)

It is worth pointing out that people who argued for science being the accumulation of facts never argued that that is all science is, just as it was usually the case that scientists who argued for hypothetico-deductivism, even those that strongly identified with Popper, argued that there is more to science than that. In this sense inductivists were quite close in opinion, though not emphasis, to those who pointed out that science is not merely the collection of facts, but also has other qualities. This, interestingly, did not come out in the books as clearly as it did in the interviews.

3.3 Science is a social enterprise

Scientists almost always acknowledged that science is a human endeavour, with all the messiness and contingencies that entailed. The influence of the social side of science on scientific knowledge was however not necessarily portrayed as negative, instead it was seen as an important and central part of science.

Most scientists held that science conforms, or should conform, to Merton-style social norms (Merton, 1973); it was often stressed that science is objective or open minded. Thirteen interviewees explicitly mentioned objectivity, eight saw science as characterised by "open-mindedness" and honesty, fifteen mentioned rigorousness and scepticism.

> what [my chemistry school teacher] used to say, "it is a scientist's bounded duty to hold his theories lightly and give them up graciously when proven else wise by somebody more ... clever". Or something like that, I mean that's the wrong words, but that's the gist of the quotation". (8 Senior, Biology, Female, UK)

This was however also often linked to an admission that this really is only an ideal, and that scientists are often fallible. Next to the comments on the rationality of science, there were plenty of admissions (eighteen) that real life science is usually much messier than that. There was a general feeling that the human side of science is very much an inseparable part of science, whether you think it is a good thing or whether you think it is regrettable but unavoidable. Even things like the personal attachment scientists have for their pet theories (mentioned seven times) is in some circumstances good thing, rather than always hindering progress as Popper would have argued:

> scientists are particularly prone to getting attached to something which objectively they might not do. And I think this is because as a scientist you need to have a good intuitive feel, and that's important for inventing hypotheses. And so, you learn to have maybe too much confidence in your intuition, which can get into the way of being objective. (23 Early Career, Physics, Male, UK)

Many more comments on the social side of science were made in the context of discussing Kuhn's philosophy later in the interviews. In these discussions the social sides of science were very often seen as necessary parts of science, and even followers of Popper agreed that Kuhn may have painted a much more realistic picture of science, warts and all.

I would argue that these topics are not necessarily the way philosophers themselves divide their subject into. On the one hand, hypothetico-deductivism is such a broad area that its several different varieties, such as falsificationism, positivism or Bayesianism were and still are fiercely contested, whereas for the scientists in my study the most important aspect of science was the basis they all shared, while the details were mostly seen as less important. On the other hand the strict division that philosophers have drawn between the empiricist/logical aspects of science and its social sides, often with the approval of science warriors, does not seem to be held widely by the scientists. Like the subjects in Gilbert and Mulkay's study, the scientists often talked about science in a "contingent" way, however (maybe because I asked them directly how that fitted in with their other views) I have heard very often that these two sides of science are both necessary for it to function, and so there was no obvious separation between the two interpretative repertoires, as Gilbert and Mulkay (1984) have found.

4 Lessons

Where scientists' representations of philosophical topics and the nature of science come from is also crucial to how scientists understand them. The education that most scientists told me they received on the nature of science was informal, based on picking up "tacit" knowledge (Polanyi, 1958; Collins, 2001), and mostly conveyed to them by other scientists. Otherwise, the context in which philosophies are picked up varied enormously from self-motivated reading, to attending lectures and even whole lecture courses on philosophy of science, run variously by scientists or philosophers. In each of these contexts philosophy gets communicated differently, and that reflects how the philosophies get interpreted.

Since scientists' representations of philosophical topics are being shaped more by other scientists themselves, and because they fulfil social psychological and rhetorical functions of identity and boundary markers, they can develop a particular dynamic. Hence, while they are certainly not naive, scientists' understanding of philosophy becomes fundamentally different to the way philosophy is understood by philosophers. The examples I elaborated in this paper illustrate the point: Popper has become an iconic figure for scientists that represents what it is to be scientific, even when he no longer commands much respect within the philosophical community. However, scientists' interpretation of what Popper stood for, and their interpretation of what it means to be Popperian is very much unlike the way philosophers interpret Popperianism. Also, when they talk about their own opinions on how science works, scientists divide the problems up in different ways from how they are discussed in philosophy, where even scientists who identify with Popper are comfortable with also holding other philosophies that are directly at odds with Popper.

For this reason—that scientists quite rationally hold several philosophies that are traditionally seen as conflicting—a simple survey of how many scientists are Popperians, and how many are Kuhnians would have given very spurious results, and it becomes clear that the way philosophy is used, thought and talked about needs to be sampled first before we can arrive at what meaningful questions to ask in the first place.

From the point of view of general science studies I believe that the sociological study of how scientists think about their activities philosophically is interesting in its own right, as it contributes to a further understanding of how science and scientists work and think about their work, and therefore elaborates on the social working of science. The philosophical reflections of scientists and how that relates to the philosophical issues debated by philosophers themselves and the discursive work philosophy performs in scientists' boundary work—these areas are generally understudied.

Studying the ways in which scientists talk about the philosophical foundations of their activity is also important from a practical view as it allows the social scientists studying science themselves to understand what scientists think are the important issues, how they are thought about and represented and what uses they are put to. This helps in understanding wider, more general, issues in science studies, as it can then inform a possible understanding of what happens when the communication between scientists and sociology of science breaks down. An example is the philosophical misunderstanding between sociologists and scientists during the science wars. Understanding the representations of philosophy by scientists can highlight where the science wars hinge on different interpretations of and significances attached to fundamental philosophical ideas. There are also some specific issues particularly in the social study of popular science which this study highlights, though this may not be a lesson as such, but rather a point to bear in mind when studying popular science. Regarding the differences between the scientists and the popular science authors in the exposition of philosophical topics, while it may in the end not be a big problem for philosophers, there are problems for popular expositions that make liberal use of philosophical asides, or philosophical topics such as Occam's razor or Popper's philosophy as a demarcation tool, or that use of philosophies as an identity marker which essentially talks to other scientists rather than the public. In light of the contested nature of philosophical topics in popular science, it is a pertinent question to ask whether the epistemologies as portrayed in popular science are giving us a consistent and realistic portrait of science.

Finally, the lesson for philosophers is that through their own social representations of philosophy, scientists have developed their own ways of talking and thinking about philosophical topics. If the formal relationship between scientists and philosophers deteriorates to the point that the philosophical opinions that scientists will inevitably have of their subject are drawn almost exclusively from among the philosophical discourse of other scientists, then philosophy of science may become slightly redundant, from a scientist's perspective. What is worse though is that in that case the philosophical discourse on philosophical topics and the scientific discourse on exactly the same topics will diverge so that the same terms and concepts will acquire different and confusing meanings. This can already be seen with reference to the different categories in which scientists think about philosophical topics in this study. There is already a movement within current philosophy of science that seeks to address the issue of philosophising on actual scientific practice, with the recent set-up of the Society for the Philosophy of Science in Practice. Although this approach addresses the relevance issue of philosophy of science, it still needs to monitor not only the way scientists practise science, but also the way science thinks about itself. Otherwise, while possibly the contents of philosophy of science can be made relevant to scientific practice, the way scientists understand that philosophy will be different to the way the philosophers understand it. Ultimately, whether philosophers like it or not, scientists will philosophise about science on their own terms, and if philosophy wants to participate in the discussion it must at least know what the philosophical issues are that scientists find relevant and interesting, but also how and in which categories they talk and understand philosophy, and when and under what circumstances philosophical topics become issues of identification and boundary work on top of their philosophical message. It is in identifying and keeping track of these types of issue that I believe sociological investigations of scientists' discourse on philosophy can make a contribution to the philosophy of science itself.

Bibliography

Abd-El-Khalick, F. and Lederman, N. G. (2000). Improving science teachers' conceptions of nature of science: A critical review of the literature. *Science Education*, 22(7):665–701.

Alters, B. J. (1997). Whose nature of science? *Journal of Research in Science Teaching*, 34:39–55.

Bailer-Jones, D. M. (2003). Scientists' thoughts on scientific models. *Perspectives on Science*, 10:275–301.

Bauer, M., Gaskell, G., and Allum, N. (2000). Quality, quantity and knowledge interests. In Bauer, M. and Gaskell, G., editors, *Qualitative Researching With Text, Image and Sound*, London. Sage.

Bauer, M. and Schoon, I. (1993). Mapping variety in public understanding of science. *Public Understanding of Science*, 2:141–55.

Brown, R. (2000). Social identity theory: Past achievements, current problems and future challenges. *European Journal of Social Psychology*, 30(6):745–778.

Brush, S. G. (1989). Prediction and theory evaluation: The case of light bending. *Science*, 246(4934):1124–1129.

Collins, H. (2001). Tacit knowledge, trust and the Q of sapphire. *Social Studies of Science*, 31(1):71–85.

Crease, R. (2001). What's your philosophy? *Physics World*, 14(10).

Crease, R. (2002). This is your philosophy. *Physics World*, 15(4):15–17.

Deutsch, D. (1997). *The fabric of reality*. Penguin, London.

Dogan, N. and Abd-El-Khalick, F. (2008). Turkish grade 10 students' and science teachers' conceptions of nature of science: A national study. *Journal of Research in Science Teaching*, 10(4):1083–112.

Donovan, A., Laudan, L., and Laudan, R., editors (1992). *Scrutinizing science*. Johns Hopkins University Press, London.

Driver, R., Leach, J., Millar, R., and Scott, P. (1996). *Young people's images of science*. Open University Press, Buckingham.

Fortey, R. (2000). *Life*. Alfred A. Knopf, New York.

Gieryn, T. F. (1995). Boundaries of science. In Jasanoff, S., Markle, G. E., Petersen, J. C., and Pinch, T., editors, *Handbook of science and technology studies*, pages 393–443, London. Sage.

Gilbert, G. N. and Mulkay, M. (1984). *Opening Pandora's box: A sociological analysis of scientists' discourse*. Cambridge University Press, Cambridge.

Hammersley, M. (1996). The relationship between qualitative and quantitative research: Paradigm loyalty versus methodological eclecticism. In Richardson, J., editor, *Handbook of qualitative research methods for psychology and the social sciences*, pages 159–74, Leicester. British Psychological Society.

Hawking, S. (2001). *The universe in a nutshell*. Random House, London.

Hogg, M. A. and Abrams, D. (1988). *Social identifications*. Routledge, London.

Huddy, L. (2001). From social to political identity: A critical examination of social identity theory. *Political Psychology*, 22(1):127–56.

Koponen, I. T. (2006). Models and modelling in physics education: A critical re-analysis of philosophical underpinnings and suggestions for revisions. *Science and Education*, 16(7–8):751–73.

Lederman, N. G. (1992). Students' and teachers' conceptions of the nature of science: A review of the research. *Journal of Research in Science Teaching*, 24(4):331–59.

Lederman, N. G. (2007). Nature of science: Past, present and future. In Abell, S. K. and Lederman, N. G., editors, *Handbook of research on science education*, Mahwah, NJ. Lawrence Erlbaum Associates.

Matthews, M. R. (1994). *Science teaching. The role of history and philosophy of science*. Routledge, London.

Matthews, M. R. (1998). In defense of modest goals when teaching about the nature of science. *Journal of Research in Science Teaching*, 35(2):161–174.

Mayr, E. (1997). *This is biology*. Harvard University Press, Cambridge MA.

Merton, R. K. (1973). *The sociology of science*. University of Chicago Press, Chicago, IL.

Miller, J. D. (1987). Scientific literacy in the United States. In Evered, D. and O'Connor, M., editors, *Communicating science to the public*, pages 19–40, Chichester. John Wiley and Sons.

Miller, J. D. (1993). Theory and measurement in the public understanding of science: A rejoinder to Bauer and Schoon. *Public Understanding of Science*, 2:235–43.

Mulkay, M. and Gilbert, G. N. (1981). Putting philosophy to work: Karl Popper's influence on scientific practice. *Philosophy of the Social Sciences*, 11(3):389–407.

Nieman, A. (2000). *The popularization of physics: Boundaries of authority and the visual culture of science*. PhD thesis, University of the West of England.

Osborne, J., Ratcliffe, M., and Duschl, R. (2003). What 'ideas-about-science' should be taught in school science? A Delphi study of the expert community. *Journal of Research in Science Teaching*, 40(7):692–720.

Polanyi, M. (1958). *Personal knowledge*. Routledge, London.

Potter, J. (1984). Testability, flexibility: Kuhnian values in scientists' discourse concerning theory choice. *Philosophy of the Social Sciences*, 14(3):303–330.

Sapolsky, R. M. (2001). *A primate's memoir*. Random House, London.

Sovacool, B. (2005). Falsification and demarcation in astronomy and cosmology. *Bulletin of Science, Technology and Society*, 25(1):53–62.

Stotz, K., Griffiths, P. E., and Knight, R. (2004). How biologists conceptualize genes: An empirical study. *Studies in the History and Philosophy of Science, Part C: Biological and Biomedical Sciences*, 35(4):647–73.

Tajfel, H. (1981). *Human groups and social categories: Studies in social psychology*. Cambridge University Press, Cambridge.

Tajfel, H. and Turner, J. C. (1986). The social identity theory of intergroup behavior. In Worchel, S. and Austin, W. G., editors, *Psychology of intergroup relations*, pages 7–24, Chicago, IL. Nelson-Hall.

Turney, J. (2001). More than just story telling: Reflecting on popular science. In Stocklmeyer, S. M., Gore, M. M., and Bryant, C., editors, *Science Communication in Theory and Practice*, Dordrecht. Kluwer.

Wilson, E. O. (1998). *Consilience*. Abacus, London.

Wong, S.-L. and Hodson, D. (2009). From the horse's mouth: What scientists say about scientific investigation and scientific knowledge. *Science Education*, 93(1):109–130.

Wynne, B. (1992). Misunderstood misunderstanding: Social identities and public uptake of science. *Public Understanding of Science*, 1:281–304.

Appendix: Popular science books from the prize shortlists

Mark Buchanan (2002). Small World. London: Orion Books.

Paul Davies, Paul (1995). About Time: Einstein's Unfinished Revolution. London: Viking.

Richard Dawkins (1995). River Out of Eden: a Darwininan View of Life. London: Weidenefeld and Nicolson.

Richard Dawkins (1997). Climbing Mount Improbable. London: Penguin.

David Deutsch (1997). The Fabric of Reality. London: Penguin.

Jared Diamond (1997). Guns, Germs and Steel. London: Vintage

Thomas Dormandy (1999). The White Death: a History of Tuberculosis. London: Hambledon.

Richard Fortey (2000). Life. New York: Alfred A Knopf.

Gerd Gigerenzer (2002). Reckoning With Risk. London: Penguin.

Steve Grand (2000). Creation: Life and How to Make It. London: Orion Books.

Brian Greene (2000). The Elegant Universe. London: Random House.

Stephen Hawking (2001). The Universe in a Nutshell. London: Random House.

David Horrobin (2001). The Madness of Adam and Eve. London: Bantam Press.

Steve Jones (1997). In the Blood: God, Genes and Destiny. London: Flamingo.

Robert P. Kirshner (2002). The Extravagant Universe. Princeton: Princeton University Press.

Armand Leroi (2003). Mutants. London: Harper Collins.

Ernst Mayr (1997). This Is Biology. Cambridge MA: Harvard University Press.

Chris McManus (2002). Right Hand, Left Hand: the Origins of Asymmetry in Brains, Bodies, Atoms and Cultures. London: Weidenfeld and Nicolson.

John Naughton (1999). A Brief History of the Future. London: Orion.

Steven Pinker (1998). How the Mind Works. London: Allen Lane.

Steven Pinker (2002). The Blank Slate. London: Penguin.

Mark Ridley (2000). Mendel's Demon. London: Orion Books.

Matt Ridley (1996). The Origins of Virtue. London: Viking.

Matt Ridley (2003). Nature Via Nurture. London: Fourth Estate.

Robert M. Sapolsky (2001). A Primate's Memoir. London: Random House.

Stephen Webb (2002). Where Is Everybody? New York: Praxis.

Robert Weinberg (1998). One Renegade Cell. London: Weidenfeld and Nicolson.

Christopher Wills (1998). Children of Prometheus. London: Allen Lane.

Edward O. Wilson (1998). Consilience. London: Abacus.

Lewis Wolpert (1999). Malignant Sadness: the Anatomy of Depression. London: Faber and Faber.

François, K., Löwe, B., Müller, T., Van Kerkhove, B., editors,
Foundations of the Formal Sciences VII
Bringing together Philosophy and Sociology of Science

Career paths in mathematics: A comparison between women and men

RENATE TOBIES

Historisches Seminar, Technische Universität Braunschweig, Schleinitzstraße 13, 38023 Braunschweig, Germany
E-mail: r.tobies@tu-bs.de

1 Introduction

In western industrialized countries, we can still find the stereotype that *Mathematics is not a subject for women*. This is reflected, for instance, in the number of female university professors: in many countries (we shall be focussing on Germany), the percentage of women among the university professors is very small. In order to discover the (historical and current) causes of this phenomenon, we analyzed the career paths of several thousand individuals who successfully completed their studies of mathematics at German universities during the 20th century. This paper introduces the resources and methods of our interdisciplinary research project and presents the main results obtained by comparing career paths of male and female mathematicians.

Some popular science books currently enjoying wide distribution suggest that women may be genetically predisposed to be poor at mathematics. These books—allegedly based on scientific results—serve to reinforce gender stereotypes and claim, among other things, that women have a weaker spatial sense than men due to structural differences of the brain. The argument goes that therefore women are less likely than men to choose professions that require spatial visualization skills. To see an example of this:

> Consulting history you will find that there are practically no outstanding women in areas that require good spatial visualization skills as well as mathematical thinking, as there is for example chess, composing music or rocket research. (Pease and Pease, 1998, p. 192)

Received by the editors: 27 October 2008; 5 July 2009; 25 April 2011.
Accepted for publication: 1 May 2011.

Yet, if we examine these claims, we discover that they are unverified;[1] it has not been proven that good spatial sense—however it is measured—is an essential prerequisite for mathematical talent.[2]

For the obvious historical reasons, female mathematicians are little known as role models. When we asked approximately 50 first-year students in our course at the beginning of the *Sommersemester 2008* to name three famous female mathematicians, the answers were alarming:[3] Only three students could name one female mathematician (Emmy Noether). Of course, this is not for lack of outstanding female mathematicians in the history of mathematics; it is particularly surprising since traditionally, mathematicians have regarded women with less prejudice than have researchers in other fields. As discussed by the sociologist Bettina Heintz (2003), this may be due to the fact that it is easier to objectively appraise performance in mathematics than in other research areas.

The famous Göttingen mathematician Felix Klein (1849–1925) suggested to his former doctoral student Johannes Schröder (1865–1937) that he investigate the mathematical instruction at public secondary girls' schools, after mathematics had finally been introduced in secondary schools in 1908. Schröder was a former girls' grammar school teacher in Hamburg;[4] one third of his students went on to study mathematics at universities. After interviewing colleagues, he concluded:

> In the past the prejudice persisted that women are completely lacking the talent for mathematical thinking. It was widely understood that their female characteristics would rather draw them towards an occupation with questions of Literature, Language, History and Ethics as opposed to the strictly logical reasoning associated with mathematics. Klein, among others, has shown convincingly how unjustified the presumption of mathematical thinking not being in a woman's nature is. (Schröder, 1913, p. 89)

In fact, Klein had supervised two doctoral theses written by female students by the year 1895. He knew that, on average, female mathematicians from abroad who attended his lectures had the same mathematical talent as their male co-students. Nevertheless, the old stereotype of women being less talented in technical subjects persisted for quite some time in most of the academic world, and especially in the humanities.

---

[1] Pease and Pease (1998, p. 166) refer to the allegedly verified surveys of the psychologists Benbow and Stanley (1983); cf. Beerman's conclusion (Beerman et al., 1992, p. 41) that this and other results of Benbow could not be verified.

[2] For the components of mathematical talent, cf. (Beerman et al., 1992, p. 15).

[3] The students were studying mathematics or some other subject at the Department of Mathematics at the *Universität des Saarlandes* in Saarbrücken.

[4] Cf. § 3.3. It was not uncommon for people with doctorates to go on to teach at secondary schools.

International statistics show that the percentage of female students studying mathematics and information technology in industrialized Western countries is smaller than in other countries. There are more female mathematics students on the Gulf peninsula and in South America than there are in Europe. In Eastern Europe, there are more women in mathematics in than in Western Europe (cf. Abele et al., 2004). Currently, Germany has one of the lowest percentages of women in mathematics. This is especially true for more senior positions in academia; the percentage of female professors in mathematics at German universities is under five percent. But what are the reasons? This question was the starting point for an extensive quantitative survey about careers in mathematics.

2 Project, methods, resources

The German *Volkswagenstiftung* supported the interdisciplinary project entitled "Women in mathematics: factors determining mathematical careers from a gender comparative perspective" initiated in 1998. We published its main results in a book with the title *Traumjob Mathematik* (Abele et al., 2004).

In our project, we followed a twofold analysis strategy: on the one hand, we collected historical and present-day data, and on the other hand we compared the historical data with the present-day data. Our objective was to analyze the careers of individuals who obtained their degree in mathematics either in the first half (historical) or at the end of the 20th century (present-day). Careers were tracked from birth to until several years after graduation using a model developed in socio-psychology due to Abele. In order to determine the factors that determine career development, we differentiated between individual-related variables (gender, origins, education and age) and external conditions (type of school system, laws).

The collection of *present-day data* is still research in progress. It deals with the careers of individuals who completed their studies in 1998. We based our research on written interviews obtained by standardised questionnaires with the most recent interview conducted in early 2008. The rate of return of questionnaires in this study was very satisfactory. Based on our experience, this indicates that people are content with their professions.

The *historiographical* research into careers in the first half of the 20th century has almost been concluded. In that time period, studies of mathematics could be completed by passing either a teachers examination or a doctoral examination; a *Diplom* in mathematics was not common before 1942. However, as early as 1920, it was possible to take a *Diplom* examination in mathematics at technical universities—but the number of graduates was low (less than 100 until 1945). We analyzed the career paths of two representative groups of people:

1. Individuals who had graduated from a teacher's course in mathematics, 1902–1940

2. Individuals who earned a doctorate in mathematics, 1907–1945

The research on the first group was based on newly discovered material from the Archive for Historical Education Research in Berlin. Over 100 boxes of personnel record cards of Prussian teachers dating from the beginning of the 19th century have been preserved; they are sorted alphabetically. We opened every third box and selected those individuals who passed their teacher examinations in mathematics.[5] We analyzed the personal records of 3,040 Prussian mathematics teachers who completed their examinations between 1902 and 1940; 462 (15.2%) of them are women.

In the second group, we considered every person who earned his or her doctorate in the reference period of these records. In 2006, we published a second book containing the short biographies of these individuals (cf. Tobies, 2006). In the course of our research, we succeeded in revealing biographical data of many formerly unknown mathematicians; among them doctoral students of Emmy Noether.

As a side effect of our research, we were able to shed more light on the position of Emmy Noether in the mathematical community. In 1998, the German mathematician Friedrich Hirzebruch stated in an interview, that "Emmy Noether did not have a school" (published as Hirzebruch, 2000, p. 1230). Our research contradicted this statement, and in response, we published the article (Koreuber and Tobies, 2002). Independently, Hirzebruch Hirzebruch (1999, p. 58) had come to the same conclusion.

We also discovered previously unknown female mathematicians who had followed job offers abroad; one of them was Cäcilie Fröhlich (Cecilie Froehlich) (1900–1992). She completed her doctorate in mathematics at the University of Bonn and became a researcher in applied mathematics at the power company *Allgemeine Elektrizitätsgesellschaft* (AEG) in Berlin. Fröhlich's career is a significant illustration of the fact that women followed careers in industrial research as early as the 1920s. However, she was Jewish and, therefore, was dismissed from her position in 1937. Even so, the AEG board provided her with an excellent certificate and she subsequently gained a position as a mathematical consultant in one of the most important power companies in Belgium (*Atelier de Constructions Électriques de Charleroi*). In 1940, she was forced to emigrate again and, finally, became a professor of applied mathematics in the electrical engineering department of the City College New York.

In Germany, the year 2008 was dedicated to mathematics as the *Jahr der Mathematik*. The *Deutsche Mathematiker-Vereinigung* celebrated the occa-

---

[5]Beginning with 1880 as a year of birth and continuing onwards from there.

sion, in part, by posting our short biographies of mathematicians on their web page. This source contains all men and women from Germany and foreign countries who completed their doctorates in mathematics at German universities from *Wintersemester 1907/08* to *Wintersemester 1944/45*, including information about their membership in the *Deutsche Mathematiker-Vereinigung*. Fröhlich's short biography can be found there.

3 Career paths of male and female mathematicians, 1900–1945: A comparison

The distribution of exams over the years clearly reflects historical context (e.g., World War I, or the congestion of schools in the mid 1920s). Our sample covering more than a third of the overall population according to Prussian statistic almanacs is representative. The distribution of doctoral degrees is similar to the distribution of the teacher examinations. Most of the graduates who earned a doctorate in mathematics also completed a teacher examination. The ratio of men and women is equal (slightly over 20%) among those individuals with successful teacher examinations who continued to pursue doctorates, though not all doctorates were in mathematics. Of all new graduates with doctorates in mathematics before 1945, only 8% were female.

Of course, there were female mathematicians earning doctorates in Germany before 1907, the beginning date for our research period; they were mostly foreign and their ambition paved the way for German women to follow suit. The first female doctor of mathematics was the Russian Sofja Kowalewskaja (1850–1891); she was awarded her degree in 1874 at the University of Göttingen and would continue her career as a mathematics professor in Stockholm. However, it would be 21 years before the next doctorate in mathematics was awarded to a woman in Germany; in 1895, Marie Gernet (supervisor: Leo Königsberger) at the University of Heidelberg, and the British Grace Chisholm (supervisor: Felix Klein) at the University of Göttingen both earned doctorates.[6]

3.1 External Conditions

Education dependent on secondary school types.

In 1908, courses in mathematics and natural sciences were offered for the first time at public secondary schools in the Kingdom of Prussia, Germany's biggest constituent state (capital: Berlin). Before the general introduction of these subjects, girls and women could acquire related knowledge in private courses only. The Kingdom of Bavaria (capital: Munich) followed suit in 1910.

---

[6] About the first foreign women who completed a doctorate in mathematics at a German university, cf. Tobies (1999) and Abele et al. (2004, Chapter 7.1).

1874. First woman to earn a doctorate in mathematics at a German university (with exemption).

1900. Permission for women to enroll at universities in the Grand Duchy of Baden.

1903. Permission for women to enroll at universities in the Kingdom of Bavaria.

1905. First woman to complete her teacher examination in mathematics (Berlin, Prussia)

1908. Permission for women to enroll at universities in the Kingdom of Prussia.

1909. Permission for women to enroll at universities in the Grand Duchy of Mecklenburg.

1919. At least until 1919, female teachers (like all state officials) had to be celibate.

1920 A law (21 February) permitted women to receive the *venia legendi* (Habilitation). In 1919, Emmy Noether already took this hurdle with exemption as the first female mathematician.

Until 1945. Only two women appointed to full professorships at German universities, none in mathematics, none of them in Prussia.

FIGURE 1. Historical overview of legal regulations.

There were four different types of secondary schools in Prussia: *Realgymnasien, Humanistische Gymnasien, Oberrealschulen* and *Oberlyzeen*. It is remarkable that most of the women in our sample completed their studies at a *Realgymnasium*. This type of school offered more lessons in mathematics and natural sciences than the *Humanistisches Gymnasium* that boys tended to attend. In the Prussian statistic yearbooks, we can find the ratio between different types of schools for girls, i.e., *Realgymnasien, Humanistische Gymnasien,* and *Oberrealschulen*. *Oberlyzeen* did not teach girls any mathematics or natural science.

Legal Regulations.

Legal regulations excluded women from academic studies and careers for a long time (cf. Figure 1 for an overview). In Germany, foreign women were the first ones to earn a doctorate, though they could only do so by exemption. The general acceptance of women for regular enrollment, teacher examinations, and postdoctoral lecture qualifications (*Habilitation* and *venia legendi*) was introduced much later in Germany than in other countries.

As a consequence of these legal regulations, only two women received a full professorship before 1945, one in the education studies at the University of Jena (Thuringia), and one in plant physiology at the *Landwirtschaftliche Hochschule Hohenheim* in Stuttgart (Wüttemberg). Moreover, until 1919, the law of celibacy forced salaried female teachers to resign from their tenure

Social background	Profession of fathers (female mathematicians)	Profession of fathers (male mathematicians)
Primary or secondary education	26.7%	20.4%
Academic education	32.0%	24.5%
Trading & business	17.2%	27.5%
Mid-level and senior civil servants	17.2%	21.3%
Military Staff	5.2%	1.3%
Working class	1.7%	4.0%

FIGURE 2. Social background of women and men who obtained their doctorate in Mathematics, 1907–1945. "Primary and secondary education" includes both teachers and education authorities. "Mid-level and senior civil servants" includes municipal, bank, post, and railway services.

upon marriage.

As soon as women were accepted for regular enrollment, mathematics evolved into a very popular field of study. The percentage of first-year female students was higher than the overall share of the female students in the 1920s. However, this changed under the Nazi dictatorship.

3.2 Personal Variables

The women covered by our sample were more likely than the men to have academically educated fathers (cf. Figure 2). Since the educations of daughters were typically funded by the parents, a studying daughter required family wealth; in contrast, sons were often allowed to study even if their families were less well off.

In our sample, the percentage of women with a Catholic confessional background (36%) is higher than the percentage of men (25.9%) and also higher than in the general population. This fact is primarily due to regional differences; Catholic areas with universities, such as Münster, Breslau, and Bonn, provided early favourable environments to women than did other places. There were several Jewish women who completed their doctorate in the first years of the 20th century and became famous researchers (e.g., Emmy Noether, Cäcilie Fröhlich) or teachers at secondary schools. Noether and Fröhlich would later emigrate but most of the secondary school teachers did not succeed in emigrating and suffered a terrible fate (cf. Tobies, 2006).

At the time of passing the teacher state examination (women $N = 462$, Men $N = 2578$), the average age of women was 27.14 years whereas the average age of men was 25.85 years. The fact that women were slightly older, corresponds to the fact that women had regularly been teaching at primary schools or medium-level secondary girls' schools before they enrolled at university.

	Women	Men
N	381	1864
Average final grade	2.17	2.21

FIGURE 3. Average final grades of teacher state examination in the years 1902–1940. The range of grades was "with distinction" (counted as 0.75), "*sehr gut*" (very good) (1.0), "*gut*" (good) (2.0), "*befriedigend*" (satisfactory) (3.0), and "*ausreichend*" (sufficient) (4.0).

3.3 Study Processes

Number of subjects in which men and women with mathematics as a main subject completed their state teacher examination

Most individuals in our sample passed their state teacher examination in three subjects without any gender difference.

Preferred Combination of subjects in the Teacher Examination

Women, like men, opted for the combination of mathematics, physics, and chemistry without any gender differences.

Universities and technical Colleges awarding doctoral degrees in mathematics, 1907–1945

The mathematicians in our sample obtained their doctorates from 23 different universities and twelve technical colleges. Based on the data, we can make the following observations:

- Göttingen was an international centre for mathematics until 1933.
- Bonn offered a particularly friendly climate for women.
- Technical colleges received the right to award doctorates later than the universities.
- At the Technical college of Dresden women were employed early and often for many years as mathematical assistants.

Teacher state examinations in mathematics: Overview of final grades

The final grades in the state teacher examinations listed in Figure 3 do not show any significant difference between men and women. Additionally, we tested a representative sample of grades for oral doctorate examinations; our findings suggest that this also holds for average grades for doctorate degrees.

Distribution of doctoral theses in mathematical subfields, men and women by comparison, 1907–1945

An older U.S. survey that compares doctoral theses awarded to men and women (from the time they were first issued until 1940) (cf. Green and LaDuke, 1987 and 2009) states that women, more often than men, wrote their dissertations in traditional fields like geometry whereas men's dissertations were more often in analysis. For instance, between 1930 and 1939, 39% of women wrote dissertations in geometry and only 25% in analysis; for men, the figures are 27% and 42% respectively.

This trend is not seen in Germany, as Figure 4 shows. Instead, the focus of research generally moved in the course of time. Until 1915, geometry was favoured by both men and women; in the following years, geometry and analysis have been chosen with equal frequency. Furthermore, the importance of applied mathematics in fields such as natural science, technology and engineering, finance and insurance mathematics, and statistics has necessitated an increase in graphical and numerical methods as well. Since 1934, the number of doctorates in applied mathematics has been equal to the amount of doctoral theses in geometry and analysis.

Generally, according to our data, there was no significant gender difference in the choice of subfield. The only significant difference that could be found was in applied mathematics; however, this difference can be explained as follows: the field of applied mathematics was expanding in the period after 1930, but it was in that period that the number of female mathematicians earning doctorates was in decline in general. We should like to emphasize, though, that a number of female mathematicians produced outstanding results in applied mathematics in such fields as aviation research, bio-statistics, and electrical engineering after they had written a thesis in one of the other fields of mathematics.

It has been a matter of debate among mathematicians and philosophers of mathematics whether you can classify mathematicians into several well-defined *types*. According to a particularly famous account, there is the *mathematical philosopher* who works from terms and concepts, the *analyst* who works with formulas, and the *geometer* who reaches new conclusions based on visualizations.[7] Our results show that differences in research style are not correlated with gender; academic work in a certain mathematical subfield employing specific methods is a result of individual education and affiliation with a particular school of thought.

---

[7] This debate was started in connection with a professorial appointment process in the year of 1892 and exclusively referred to men; in the debate, some famous mathematicians were classified according to these three types. The conceptual thinking approach has been largely attributed to Richard Dedekind and Georg Cantor. However, Emmy Noether and her students subsequently worked in the same fashion.

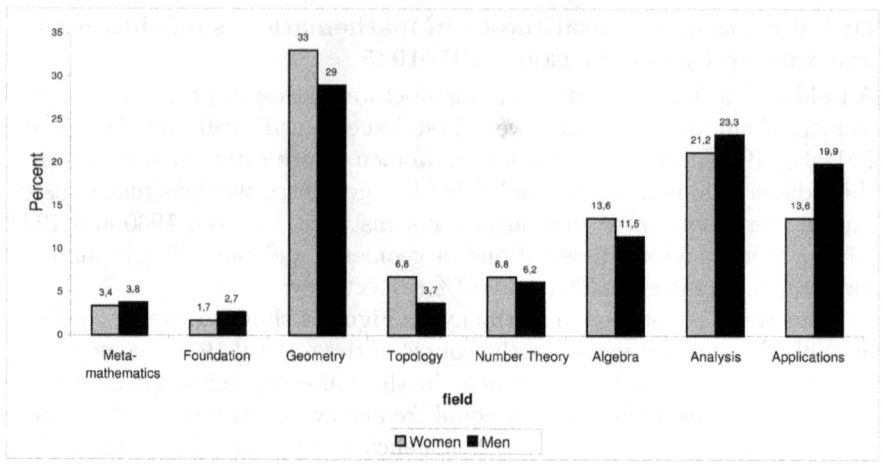

FIGURE 4. Distribution of doctoral theses in mathematical subfields at German universities, 1907–1945

Before the 1930s, secondary school teaching was the most common career path for mathematicians, even those with doctorates; this changed, however, when the world-wide economic crisis caused a hiring freeze in secondary schools. At the same time, mathematically educated people were widely recruited and promoted by the business and military industrial complex in connection with the war effort; significant examples are the aircraft industry and aviation research. Here also women found employment—for example, in rocket research, where we discovered a number of outstanding female scientists.

Here, Irmgard Flügge-Lotz (1903–1974) and Ingeborg Ginzel (1904–1966) were important pioneers (cf. Tobies, 2008); they were awarded doctorates at the end of the 1920s at the *Technischen Hochschulen* in Hannover and Dresden, respectively. After that, they worked in aviation research at the *Aerodynamische Versuchsanstalt* in Göttingen. Immediately after the war, it was not possible to continue in aviation research in Germany, so they continued their careers abroad, as did many of their male colleagues. Flügge-Lotz went to Stanford University via France and Ginzel went to the U.S. aircraft industry via the United Kingdom. Ginzel worked with the company that manufactured the bombs dropped on Nagasaki and Hiroshima, one of the largest companies in the U.S. market. This clearly demonstrates that women are not antimilitaristic by nature, just as they cannot be said to be predisposed to specific characteristics or have specific academic subject preferences.

4 Historical and present-day careers: A comparison

Women historically favoured mathematics as a subject of study and they still do. Women currently represent over 50% of those studying mathematics for a teaching profession. In Germany, women studying mathematics generally become teachers. Decisions of this type are often related to the fact that a state school teaching post allows the combination of a professional career and family life in a way that is impossible when working for a private employer. Women account for approximately 40% of those enrolled in *Diplom* programmes.

In summary, our main findings are that historically, the socioeconomic background of female students was distinctly different from that of male students; today that is no longer the case. The career entry for men and women was and still is similar. In the early 20th century, there as a great deal of legal discrimination against women in mathematics; today the main challenge that women face is balancing career and family life. Neither historically nor presently, we can find any gender-related differences in performance parameters such as final degrees or duration of study. However, differences do exist between students studying to become teachers and those studying to receive the *Diplom* degree. On average, *Diplom* students earn higher grades but study longer. Female students pursuing teaching positions are the fastest to finish their studies; a random sample of historical data shows an average length of study of nine semesters. In the past, studies were finished considerably faster, even though the place of study was changed more often and examinations were normally taken in three subjects.

Upon beginning a career, there is no difference between men and women; even three years after their exams, there are no significant disparities. Over 50% of the individuals that passed the teacher examination are now working in their profession; others are still in training. Of those who are unemployed, 44% are mothers.

Shortly after their examinations, there is no gender disparity among *Diplom* graduates either; currently, most find employment regardless of gender if they leave academia after the *Diplom* degree. Among doctoral students, there is no gender disparity in the number of total years spent at the university. Five years after graduation, men and women have the same likelihood (13%) of having reached the status of supervisor with at least six employees under them. However, this is not true of women who have become mothers.

In conclusion, we can stress that the successful integration of personal and professional life is an important component to personal satisfaction.

5 Concluding remarks

Before 1945, there was no difference in the career paths of women and men up to the position of a *Studienassessor*. Due to the legal situation, in order to obtain higher positions, women had to choose career over family. Today, female mathematicians without children reach the same positions and salaries as their male colleagues. At the same time, there is a great disparity if we consider the total number of women in senior positions in mathematics.

This raises the following natural questions: What are the causes of this disparity? There is a great number of female students and graduates, but where do they go and why? (In mathematics as in the humanities the vast majority of full professorships are held by men.) Are women generally less career-oriented than men? Is this related to the decision making process for hiring of professors? Who is part of this process and how do they affect it? What are the current options of combining career and family? Is the decision against children that we observe in many of the successful female mathematicians a consequence of an internalized "uncaring mother" prejudice? (This seems to be a specifically German concern.)

Still, women are more likely than men to end, freeze or slow down their careers for the benefit of their families. In 2006, 89.3% of interviewed female mathematicians reduced their professional engagement in favour of children, for men it was only 28.45%. Our results show that men are, in general, more career-oriented than women. There is a strong correlation between career-orientation before graduation and success in the profession after graduation.

A large number of the interviewed women prefer a so-called *soft career*, which includes family phases (marriage, children, etc.) from the outset. As a consequence, in Germany, academically educated women regularly stay home for a considerable time after giving birth. On one hand this is due to insufficient child care facilities, on the other hand, the so-called "uncaring mother" argument looms ever-presently in the backs of minds. Very importantly, the interviews revealed that many women perceive slow professional advancement as a sign of having successfully integrated family and professional life.

It is beyond doubt that science as a whole is suffering if talented researchers are not doing research. Many excellent female mathematicians decide to quit research after the doctorate. Among other reasons, they name the required mobility for junior researcher that is a necessity for a scientific career in Germany. So, in order to increase the number of women achieving leading positions in mathematics, we must work on the integration of career and family and raise the awareness among female researchers that careers in research are possible, can, and have to be planned.

1988. *Abitur*, Frankenthal (Pfalz).

Studies of mathematics with minor in business studies at the University of Kaiserslautern.

1994. *Diplom* in mathematics.

1998. Doctorate in mathematics (Kaiserslautern).

2003. *Habilitation* (Kaiserslautern) 2003.

2004. Professorship (W2) at the Institute for Numerical and Applied Mathematics, University of Göttingen.

FIGURE 5. Career milestones of Prof. Dr. Anita Schöbel

In fact, there are already good role models available who exemplify how women can integrate a professional and personal life; one is Anita Schöbel who is married with two children. In Figure 5 is a list of her career milestones. The successive longitudinal study will clarify whether the 21st century will produce more successful examples like Professor Schöbel.

Bibliography

Abele, A., Neunzert, H., and Tobies, R. (2004). *Traumjob Mathematik! Berufswege von Frauen und Männern*. Birkhäuser, Basel.

Beerman, L., Heller, K. A., and Menacher, P., editors (1992). *Mathe: nichts für Mädchen? Begabung und Geschlecht am Beispiel von Mathematik, Naturwissenschaft und Technik*. Hans Huber Verlag, Bern.

Benbow, P. C. and Stanley, J. C. (1983). Sex differences in mathematical reasoning ability: More facts. *Science*, 222(4627):1029–1031.

Green, J. and LaDuke, J. (1987). Women in the American mathematical community: The pre-1940 Ph.D.'s. *Mathematical Intelligencer*, 9(1):11–23.

Green, J. and LaDuke, J. (2009). *Pioneering women in American mathematics. The pre-1940 PhD's*, volume 34 of *History of Mathematics*. American Mathematical Society, Providence.

Heintz, B. (2003). Die Objektivität der Wissenschaft und die Partikularität des Geschlechts. Geschlechterunterschiede im disziplinären Vergleich. In Wobbe, T., editor, *Zwischen Vorderbühne und Hinterbühne. Beiträge zum Wandel der Geschlechterbeziehungen in der Wissenschaft vom 17. Jahrhundert bis zur Gegenwart*, pages 211–237, Bielefeld. transcript Verlag.

Hirzebruch, F. (1999). Emmy Noether and topology. In Teicher, M., editor, *The heritage of Emmy Noether. Proceedings of the conference held at Bar-Ilan University, Ramat-Gan, December 2–3, 1996*, volume 12 of *Israel Mathematical Conference Proceedings*, pages 57–65, Ramat Gan. Bar-Ilan University.

Hirzebruch, F. (2000). An interview. In Pier, J.-P., editor, *Development of mathematics 1950–2000*, pages 1229–1239, Basel. Birkhäuser.

Koreuber, M. and Tobies, R. (2002). Emmy Noether – Begründerin einer mathematischen Schule. *Mitteilungen der Deutschen Mathematiker-Vereinigung*, 3:24–37.

Pease, A. and Pease, B. (1998). *Why men don't listen & women can't read maps: how we're different and what to do about it*. Pease Training International, Mona Vale. Page numbers refer to the German translation (Pease and Pease, 2000).

Pease, A. and Pease, B. (2000). *Warum Männer nicht zuhören und Frauen schlecht einparken. Ganz natürliche Erklärungen für eigentlich unerklärliche Schwächen*. Ullstein Taschenbuch, München.

Schröder, J. (1913). *Die neuzeitliche Entwicklung des mathematischen Unterrichts an den höheren Mädchenschulen Deutschlands insbesondere Norddeutschlands*, volume 1 of *Abhandlungen über den mathematischen Unterricht in Deutschland*. Leipzig, Teubner.

Tobies, R. (1999). Felix Klein und David Hilbert als Förderer von Frauen in der Mathematik. *Prague Studies in the History of Science and Technology*, N.S., 3:69–101.

Tobies, R. (2006). *Biographisches Lexikon in Mathematik promovierter Personen*, volume 58 of *Algorismus, Studien zur Geschichte der Mathematik und der Naturwissenschaften*. Dr. Erwin Rauner Verlag, Augsburg.

Tobies, R. (2008). *"Aller Männerkultur zum Trotz": Frauen in Mathematik, Naturwissenschaften und Technik*. Campus, Frankfurt a.M.

 François, K., Löwe, B., Müller, T., Van Kerkhove, B., editors,
Foundations of the Formal Sciences VII
Bringing together Philosophy and Sociology of Science

Alternative claims to the discovery of modern logic: Coincidences and diversification

PAUL ZICHE*

Departement Wijsbegeerte, Universiteit Utrecht, Janskerkhof 13a, 3512 BL Utrecht, The Netherlands
E-mail: Paul.Ziche@phil.uu.nl

1 Multiple and alternative discoveries

No discipline could lay stronger claim to clarity and unequivocality than logic. Curiously, though, the historical genesis of modern logic presents a picture riven with rival claims over the discipline's founding contributions. As will be shown, protagonists from highly different backgrounds assert that the genesis of modern logic—indeed, its very discovery—rests on their contribution. Not only are these claims, to some extent, mutually exclusive, they also cut across standards of scientificity and rationality. By standard narratives of the history of modern logic, some claimants to paternity seem downright obscure and anti-rational. Yet, all these claims are made within a narrow time-frame, and they all refer back to the same developments in mathematics. This makes it impossible to easily dismiss the rival claims.

This paper argues that the co-existence of alternative claims concerning the discovery of modern logic, in fact, places the historian in an advantageous situation. It might appear as if the alternative claims make the the genesis of modern logic into a story of a *multiple discovery*, the coincidence of simultaneous discoveries of the same phenomenon. This is, however, not what the historical picture shows. The competing claims, very explicitly placed by the protagonists themselves, rather have to be viewed as stages within an open-ended process that would contribute not to an integrated picture of logic, but generate diversified conceptions of rationality. It is

---

*Many thanks to the referees, to Thomas Müller and in particular to Cheryce Kramer for improving my text in many ways. The two illustrations from (Mannoury, 1934) are reproduced with kind permission of Springer-Verlag.

Received by the editors: 6 May 2009; 7 March 2010.
Accepted for publication: 3 August 2010.

precisely the differences between the positions that tended to be emphasized. Fundamental notions, seemingly incontestable today, were still very much in dispute. These disputes as to the discovery of modern logic can serve to both enrich our understanding of the historical milieu in which logic emerges as a modern discipline and, perhaps more importantly, question extant normative standards used to reconstruct the historical genesis of logic.

Multiple discoveries are, at least since Thomas Kuhn (1959) and Robert K. Merton (1961, 1963), a well-established topos in the historical and philosophical reconstruction of scientific discoveries. They are usually viewed as two (or more) distinct and independent discoveries of the same result (which may pertain to any stock element of science, such as theories, laws, relevant observations etc.). In many cases—the standard example being the multiple discovery of the principle of energy conservation—these discoveries are later integrated into a comprehensive theory, thereby revealing the various contributions as being only partial discoveries. Serious problems, however, hamper this account: in most cases, it is very difficult to determine whether or not the discoveries are really totally independent. Can it really be the case that two researchers working within the same period can be considered completely independent? Likewise, it is hard to establish the identity of the discovery. What does it mean for two discoveries that are made in different contexts to be identical in all relevant aspects? Kuhn posed this question in his paradigmatic paper on the discovery of energy conservation without, however, exhausting its implications in detail. Strictly speaking, a simultaneous discovery would presuppose that different protagonists announced "the same thing at the same time" but that is not what one actually finds (Kuhn, 1959, p. 70). Far from being a well-orchestrated set of simultaneous discoveries, the historical process of discovery can rather be seen as a series of partial events that require further integration and become identified or unified only in hindsight.

The situation changes in important respects when switching to the scenario that shall be dubbed *"alternative discoveries"*. These are cases where, starting from what may appear as a multiple discovery, different scientists explicitly claim to be the first, or the genuine, discoverer of a relevant result. The multiplicity is thus seen and stated not retrospectively by historians, but by the protagonists themselves. What sets cases of alternative discovery apart from multiples, in particular, is the aim behind stating the multiple discovery. When a scientist actually claims to have contributed to a multiple discovery, this may be just a pitch for priority. In many cases, however, a different goal is pursued: the alternative claims tend to emphasize precisely the uniqueness, the singular value of the various contributions, respectively, and thus focus on the *differences* that lie behind seemingly very similar or even identical discoveries. This is especially revealing in disputes concerning conceptual aspects of the various discoveries. Lines of differentiation are, for instance, grasping the true significance of a discovery, giving it its proper place within a larger theory, embedding it in a general context, giving the best proof, showing the most salient application. Alternative discoveries, in this sense, stand between the identity typically generated in cases of multiple discoveries, and the emphasis on differences between the various partial discoveries. This emphasis also reveals where the protagonists themselves located the most relevant innovations.

The alternative claims regarding the discovery of modern logic cannot be integrated smoothly into one great multiple discovery. That logic is, in any case, intimately related with rationality and with the foundations of "science" (in the broad sense) remains uncontroversial. Yet, on the alternative discoveries' reading, differences become visible that reveal that even in fundamental questions of modern logic the bandwidth of possible options was far broader than one might expect. Since logic pertains to foundational ideas, this suggests that even among research traditions that share a common ideal of scientific rationality, this ideal could be understood in rather different ways. The relevant question, then, is not: have the actors involved indeed made the very same discovery but, rather, where do the relevant lines of agreement and disagreement lie and to what extent can this mapping inform our understanding of fundamental notions at play in the 'official' account of the genesis of modern logic?[1]

2 Alternative claims: contesting the discovery of modern logic

Around 1900, authors from highly different, even—from today's perspective—absolutely irreconcilable traditions lay claim to the discovery of modern logic. The following features are taken to be defining characteristics of 'modern' logic: the new logic is based on mathematics and takes novel developments in 19th mathematics into account; it incorporates a theory of relations and thus goes beyond traditional syllogistics; it aims at developing a type of science that occupies the *most fundamental* place in the system of scientific disciplines.

Yet, these alternative claims have not become part of the common historiography of logic. Here, typically, a mathematical-philosophical line is

---

[1] Pulkkinen (2005) illustrates the difficulties inherent in singling out an 'official' line of the development of logic in the 19th century. Peckhaus's highly critical review (Peckhaus, 1997) on the one hand draws a yet richer picture, while on the other hand trying to clarify the situation further. For an appreciative account of some of the alternative lines in the history of logic, cf. also Bowne (1966); Hansen (2000); Haaparanta (2009); the latter is a broad overview still based on the official history of logic. For a case study concerning multiple discoveries or re-discoveries in the field of logic, cf. Schlimm (2011).

singled out and has become the 'official' lineage, with well-known and generally respected protagonists such as de Morgan, Boole, Peano, Frege, Russell or Whitehead, and with a number of lesser known, but nonetheless equally accepted players such as Louis Couturat. Other authors, associated with more traditional philosophical projects (such as Sigwart, Wundt or Lotze), were largely eclipsed by changes within the field of philosophy, and by the increasing emphasis on the role of mathematics for modern logic.[2] The rivals who laid alternative claims as to the discovery of modern logic are known neither as possible claimants for the title of an inventor of modern logic, nor are they considered part of the official history of philosophy; they are simply banished to obscurity. Perhaps the most prominent claimants in this context are Wilhelm Ostwald (1853–1932), the Nobel Prize winning Leipzig chemist, founder of the discipline of physical chemistry and popularizer of a "monistic" world-view, and Hans Driesch (1867–1941), pupil of the equally famous and infamous zoologist and *Weltanschauungs*-thinker Ernst Haeckel, an embryologist of the first rank whose philosophical ideas on irreducibly a-mechanistic "entelechies" in living organisms were presented by Vienna Circle members as the very paradigm for anti-scientific metaphysics (Carnap et al., 1929, p. 312).[3]

Although the recent literature has devoted considerable attention to the mutual influences and points of intersection between different traditions in 19th and 20th century philosophy such as neo-Kantianism, phenomenology and analytical philosophy,[4] the emergent picture remains pointillist; it focuses on individual terms or concepts and excludes awkward alternative thinkers such as Ostwald or Driesch. Yet, the lines separating these more radically alternative traditions from 'official' analytical philosophy are by no means sharp: Ostwald and Russell pursue at least one common goal by publishing, in German and English, respectively, Wittgenstein's *Tractatus*; Ostwald and Couturat join forces in propagating new artificial languages (such as Esperanto or, preferred by Ostwald and Couturat, "Ido") for daily use (cf. Ziche, 2009). And in one of the early classics of analytical phi-

---

[2] Yet other players such as Husserl and Peirce are perhaps even more difficult to place; they will not be discussed here.

[3] On Driesch as biologist, cf., e.g., Jahn (1998, pp. 444–445); on Ostwald, cf., e.g., Görs et al. (2005).

[4] Important contributions come from Gottfried Gabriel, Hans Sluga, Michael Friedman and Alan Richardson. An example for the pointillist character of many studies of the interactions between the fields is Sluga (1997). When discussing Frege's indebtedness to the Neo-Kantians that is visible in his use of the term "Wahrheitswert" (a term playing a considerable role also in the "significal" debates discussed in this paper in §§III and IV), Sluga describes the relationship between Frege and his Neo-Kantian contemporaries in terms of a "quotation", a "borrowing", a "link" and of "connections", but stresses that "[i]t is precisely in the nature of such borrowings that they are partial" (Sluga, 1997, pp. 31–32).

losophy, Carnap's *Logischer Aufbau* from 1928, Ostwald and Driesch stand peacefully next to Russell and Whitehead (Carnap, 1928, pp. 3–4).[5]

Ostwald bases his scientific world-view on one single principle—in other words, he argues vigorously for a 'monism'—, namely the principle of energy conservation. Although he duly celebrates the innovative character of this principle and the heroic achievements of its 19th-century discoverers, he replaces it in its function as the most general principle in science by another principle that he claims to have introduced himself. When rewriting his *Lectures on the Philosophy of Nature* from 1902 for a second edition appearing in 1914, he states that it was him, Ostwald, who had discovered (with the disclaimer "to the best of my knowledge" whose modesty appears, given his repeated statements to the same effect, as rather tongue-in-cheek) that a great progress in the philosophy of science can be achieved by acknowledging "logic as the first and most general science ['Wissenschaft'], even more fundamental than mathematics" (Ostwald, 1914, VI). Already in 1909 he had stated, referring in very concrete terms to the mathematical theories forming the inspiration for his claim, that he had made the important discovery "that logic, which better and more generally should be called the theory of manifolds, is an even more general science than mathematics" (Ostwald, 1909).[6] With his reference to theories of manifolds, he refers to a mathematical concept that had, in different meanings but always with highly innovative results, been used by Hermann Grassmann in his multi-dimensional algebra, by Bernhard Riemann in differential geometry and by Georg Cantor in his set theory, and that had provided concepts that proved indispensable in the genesis of modern logic. An admirable overview over these theories is given in Whitehead's first major publication, his *Treatise on universal algebra with applications* (Whitehead, 1898).

What is the basis of these claims as regards logic? Ostwald links his interest in mathematics and logic to the problem of concept formation and the task of constructing new types of languages for the purpose of absolutely clear communication. Without mentioning Frege,[7] he adopts the term "Begriffsschrift" for his own program of establishing one (or several) languages that aspire to complete definiteness. For him, the new logic devolves from the attempt to generalize one of the central tasks of science, that of clarifying

---

[5] Driesch is mentioned here, under the title "Konstitutionstheorie"—a term that Carnap employs to characterize his own project—as the author of one of the most important modern "Begriffssysteme"; Ostwald is seen as a more distant point of reference because Ostwald (just as the psychologist-philosophers Wundt and Oswald Külpe or the theologian Paul Tillich) gives a classification of forms of science, i.e., of systems of concepts, but no derivation of the concepts employed in science.

[6] On the broader context, cf. Ziche (2006, 2008).

[7] It seems clear that Ostwald came into contact with Fregean ideas in the context of publishing Wittgenstein's *Tractatus*; however, there seems to be no explicit reference to Frege's work in Ostwald's writings.

our concepts so as to arrive at well-defined concepts. Just as in Helmholtz's *Thatsachen in der Wahrnehmung*, and in the principle of energy conservation, this procedure is substantiated by the fundamental role played by invariants and relations in science (von Helmholtz, 1878).[8] In new algebraic theories, such as the theory of groups or of manifolds, Ostwald discerns a general framework for thinking order and structure in terms that can be applied to concept formation.

The same terms are featured in Hans Driesch's "Ordnungslehre". Driesch views his "Ordnungslehre" explicitly as "logic, understood in the broadest sense" (Driesch, 1923, p. 2).[9] He does not talk in terms of a discovery of modern logic, but in terms of giving logic its proper place within the system of various sciences. His "Ordnungslehre" aims to rival the status of logic as the most fundamental science: "Ordnungslehre" is "die wahre erste Philosophie", the genuine first philosophy (Driesch, 1913, V). Interesting enough, he mentions in one and the same paragraph Ostwald (critically, because he does not really penetrate to the level of genuine metaphysics) and, more approvingly, Moritz Schlick (because he at least makes definite statements regarding the inexistence of an empirical reality) (Driesch, 1926, p. 8).

Some common patterns, and some marked differences between the different origin stories of logic, stand out immediately. All protagonists, regardless of whether they had a formal education in mathematics or not, looked at very much the same theories within mathematics to substantiate their claim that a "new logic" was both required and possible. The two most important examples are new algebraic theories such as group theory, generalizing on the usual forms of algebraic operations, and the theory of manifolds as a natural extension of traditional theories within arithmetic and geometry. These developments are intimately linked with the re-assessment of the seemingly most obvious foundations and fundamental operations of mathematics. A particularly prominent example for the interplay between highly innovative achievements and very basic topics is a new interest in the

---

[8]On the history of the logic of relations, cf., e.g., Merrill (1990); on invariants as a mathematico-philosophical issue, cf. Ihmig (1997).

[9]His "Ordnungslehre" starts with a—possibly solipsistic—immanentism, and is intended to prepare the ground for any future metaphysics. The fact that I experience something that is ordered in a particular way ("Erleben von bestimmtem Geordneten", Driesch, 1923, p. 1) forms the basic topic of philosophy. References to recent results in mathematics are to be found, e.g., in the chapter on "manifold" (p. 136) where Driesch refers to Riemann—but only in order to distinguish his concept of manifold from the corresponding concept in mathematics. Driesch also employs his concept of "order" in his works on theoretical biology, e.g., in the programmatic treatise (Driesch, 1924), and throughout his Gifford lectures on the philosophy of the organic world (Driesch, 1909). Cf. also Vollenhoven (1921).

justification of the complex numbers.[10] Here, however, the participant actors already part company: shared interest in these mathematical theories and a shared agenda to develop new and more precise forms of language no longer produce the same conclusions. The cases of Ostwald and Driesch indicate that, in taking up inspirations from recent trends in mathematics, one could perfectly well focus on very simple cases of application (such as complex numbers) and still think that one had incorporated the most essential aspects of these mathematical innovations. Formal language, then, becomes dispensable.

It is also striking that the typical career paths of the protagonists involve a high degree of switching between fields and disciplines, be it mathematics and philosophy, or, in the case of Ostwald and Driesch, philosophy and the natural sciences. This makes sense in a context where the stated task is disciplinary innovation on both ends of the spectrum of practice: at the level of maximum generality and at the level of highest speciality. Driesch and Ostwald were, in one respect, the most radical among the group of discoverers of logic: they opted for a change of the name of the fundamental discipline they aimed at, based precisely on the new achievements within mathematics, suggesting terms such as "Mannigfaltigkeitslehre" or "Ordnungslehre" for the new discipline. Debates on logic thus fit into the heyday of creating new disciplines; "Gegenstandstheorie" (Meinong, but also Driesch), "Phänomenologie", but the same holds for "Ordnungslehre"—which then should be compared with the role the concept of order comes to play in, for instance, Whiteheadian philosophy of nature.

3 Synthesis in *Synthese*: Pluralistic origins of a logico-philosophical journal

The claims made by Ostwald and Driesch are more than just exotic (or quixotic) hobby-horses of German philosophers of nature. Precisely the same discursive formations can be found in some of the most relevant philosophical journals of this time, and it is in these journals that the rather abrupt changes in conceptions of rationality show most clearly. The very form of these publications typically combines programmatic statements with a diversity of contributions that normally cover a considerably broader range. Even journals that function as the mouthpiece of a rather well-

---

[10]"New" is here taken in the sense that in many publications from highly divergent fields (mathematics, logic, literature, philosophy...) from the years around 1900 one finds a strong tendency to view the complex numbers (despite older discussions in Gauss or Riemann, among others, and despite the well-established use made of complex numbers in the natural sciences) as "paradoxical" or "contradictory". Examples are, among philosophers and philosopher-mathematicians, Husserl (1901, p. 433) and Natorp (1923, p. 239), among mathematicians and logicians, Hankel (1867, V). Cf. Ziche (2008, Chapter VI.3).

defined group of authors typically present a high degree of internal diversity. The alternative traditions in the discovery of modern logic figure alongside each other in one of the first issues of a journal that was to become one of the central forums for modern analytical philosophy, *Synthese*. Here we find rather extensive and affirmative discussions of the role of Driesch as philosopher but also a marked interest in the ideas of Moritz Schlick as well as a whole series of articles by the ex-theosophist Matthieu Schoenmakers.[11] Similar observations can be made with regard to other journals such as *Erkenntnis* or *Mind*.[12]

The early history of *Synthese* offers a telling case study for present purposes. *Synthese* started as the official organ of the "signifische kring", a group of Dutch intellectuals that perfectly illustrates the problems encountered in the attempt to give a unilinear account of the genesis of modern logic or of analytical philosophy. The movement of "significs", based on the philosophy of language of Lady Victoria Welby, was a predominantly Dutch affair that attracted thinkers with rather different backgrounds. Prominent members of this movement were the writer Frederik van Eeden, the mathematician and philosopher Gerrit Mannoury, the great mathematician and logician Brouwer and the linguist Jacob van Ginneken.[13] The official "Introduction" to *Synthese* takes up basic ideas of this group: the journal aims to express the spirit of the day, "de geest van onzen tijd", which was conceived as a "period of synthesis". A deeply and widely felt longing for a synthesis had been unleashed by the increasing specialization and fragmentation that typified the 19th century. A broad alliance is formed, across schools, traditions and disciplines. In the eyes of the editorial board of *Synthese*, "eminent biologists" fight together with Hegelians for the unity of analysis and synthesis. Categories from the philosophy of the human sciences such as "gelden", i.e., "to have a value", are employed next to motives from gestalt psychology, and it would be hard to conjure up a more diversified list of reference authors than those cited by the editors of *Synthese*: the psycho-physicist Fechner and the physicists Jeans and Einstein, the latter for contributing to a "crisis der zekerheden", a "crisis of certainty"; the mathematician-philosophers Henri Poincaré and the more heterodox Émile Boutroux (his

---

[11] Cf., e.g., his articles on "Zinzeggende beeldspraak" (roughly: a form of image-based or image-directed language that expresses the essence of reality), with the thesis that "Ook de exacte formule is beeldsprak" ("the exact formula, too, is figurative speech") (Schoenmaekers, 1936, p. 7). On Schoenmaekers, cf. de Jager (1992).

[12] On the early history of *Erkenntnis*, cf. Hegselmann and Siegwart (1991); they show how *Erkenntnis* emerged from a strong neo-Kantian background.

[13] The history of the significal movement cannot be presented here; a comprehensive overview is to be found in Schmitz (1990). Cf. also the extensive commentary on one of the early texts of the significs-movement (van Eeden, 1897). For the broader context and forms of institutionalization, cf. Heijerman and van den Hoven (1986). On Brouwer and the significal movement, cf. van Dalen (1999, esp. pp. 243–250, 255–270, 367–375).

writings include texts on mathematics and mysticism, and on the philosophy of German Idealism); the idealist Jean-Marie Guyau and the pessimistic and/or metaphysical authors Schopenhauer, van Hartmann, Nietzsche and Bergson. Equally diversified is the list of psychologists/physiologists that *Synthese* references, including von Monakow, Stern, van Maeder, Jung and Janet (Godefroy et al., 1936b).

The motive of a crisis remains prevalent and leads to a whole range of strongly negative statements: "our time is in essence anti-materialist, anti-intellectualist, anti-mechanistic". Synthese, however, does not recommend a sceptical or pessimistic metaphysics but aims at integrating all these tendencies into a larger unity modelled on the scientific mode of thinking. This, in turn, is described as "an orientation towards a clearer and more powerful unity" ("oriëntering naar klaarder en krachtiger eenheid"; Godefroy et al., 1936b, p. 6). This formulation shows how different sets of categories coexist at the time in one and the same journal. "Clarity" is an ideal that stands central also in the early manifestos of the Vienna Circle and of analytical philosophy, the terms "powerful" or "forceful" thinking, however, sooner evoke the ideas of metaphysically minded authors such as Nietzsche, Schopenhauer or von Hartmann.

Most revealing is the way in which the significal authors treat the Vienna Circle in *Synthese*. Much space is devoted to Moritz Schlick, in particular.[14] Already in the first volume of *Synthese*, Schlick is viewed as a thinker who brings several central issues and problems of current-day thinking into sharp focus (Godefroy et al., 1936a, p. 108). The journal quotes extensively from his writings, and at least one passage expresses strong approval: "the essential and highly pleasing aspect, the living modernity in Schlick is his spiritual drive ['geestesdrang'] towards a synthesis of philosophy and the special sciences ['vakwetenschap']" (Godefroy et al., 1936a, p. 112). In later publications, for instance in an obituary devoted to Schlick in 1936, Schlick is characterized even more explicitly in world-view terminology. He is celebrated as a philosopher who reflects on the "essence" ("zin") of his own time, and who thereby tries to bring about an integration of all areas of knowledge, a unity of science based on strictly scientific principles (Godefroy et al., 1936c, p. 195). This unity also implies a unity of the modes of expression in the sciences. When the editorial board of *Synthese* states that "what our times desperately needs ['broodnodig'] are clear and transparent concepts" (Godefroy et al., 1936c, p. 197), they state both their own and Schlick's goal. Again, they combine categories and forms of expression belonging to a *Weltanschauungs*-discourse with those stemming from the official tradition of analytical philosophy.

---

[14] After the war, Otto Neurath (who also was a member of the "Internationale signifische studiegroep", cf. Schmitz, 1990, p. 18) publishes a programmatic overview over the unity-of-science-movement (Neurath, 1946).

Still, there are marked differences between the Vienna Circle and the significs, and these are rooted in fundamental tenets of their respective beliefs. These are clearly formulated in an article that discusses the contrast between "Significa" and the Vienna Circle (Godefroy et al., 1936d). Here, the program of bringing "clarity" into our diffuse and chaotic ("vertroebeld") emotional and intellectual life again serves, from the very first sentence on, as the leitmotif. Language is identified as the cause of this confusion. Yet, the Dutch and the Viennese approach to language differ: in Vienna-style thinking, it is the language of science and of philosophy that comes under scrutiny, whereas in the Netherlands, it is ordinary everyday language that the significal thinkers scrutinize (Godefroy et al., 1936d, p. 332). Again, however, in concrete practice there is a close agreement. The Dutch thinkers propose to work on a large dictionary/encyclopedia-project that should capture the fundamental words ("grondwoorden") necessary for any successful use of language. On the basis of such a fundament, the different levels of language ("taaltrappen", literally: steps of language) could be reconstructed that mediate between the fundamentals of everyday language and the use of language in logic, mathematics or science.

The idea that there are fundamentally different forms of language, related via a whole series of steps, is central for the significs' approach. The focus on language, however, leads to rather sharp divergences between the groups of authors represented in *Synthese*. This becomes particularly clear in another joint issue that interests both the significal authors and the Vienna Circle: the adequate way of dealing with "Scheinprobleme", i.e., with pseudo-problems in science and philosophy. One way of putting the latter problem distinguishes between "formal-logical" and "immediate-intuitive" uses of language as two extremes that, nevertheless, belong to a unity of language that is to become the subject of significal research (Godefroy et al., 1936d, p. 336). The intuitive, and thereby also the affective and emotional aspects of language are a fundamental and irreducible element of language; "significa comprises more than just a critique of language, or even a synthesis of language; significs has to be directed towards a deeper understanding in the connection between words and contents of the soul ['zielsinhoud']" (Godefroy et al., 1936d, p. 336–337). Still, *Synthese* aims at a grand synthesis, based on significs and on the results of the Vienna Circle—more precisely, on the "experiences" made by them—with the intention of bringing "some order into the chaos of the thinking of today". That the editors of *Synthese* could hold that such an approach, based on approaches they themselves took to be highly diversified, could establish order and clarity shows that the concept of clarity itself is far from clear (Godefroy et al., 1936d, p. 338).

During and after the Second World War, some more explicit statements are formulated with regard to central ideas of the thinkers of the Vienna circle, beginning with the question of "Scheinprobleme" and amounting to a differentiation of uses of logic. Meaningless expressions can occur in metaphysics, "[b]ut we do hold the opinion that a logical analysis does not suffice to brand as pseudo-questions, meaningless questions and 'isolated sentences' the many questions appearing in diverse 'doctrines regarding views of life'" (Clay et al., 1946a, p. 22).[15] Again, a highly interesting and highly diversified list of authors is quoted as supporting the significs' investigations into the foundations of knowledge: "Husserl, Mach, Hans Vaihinger, Driesch, Carnap, Schlick" (Clay et al., 1946a, p. 22).[16]

The inclusive attitude lasts surprisingly long. The introduction to the post-war issue from 1946 likewise stresses the conciliatory character: "we tried to avoid going to extremes" (Clay et al., 1946b, p. 9), steering a middle way between the forms of "word-thinking" predominant "in the off-shoots of Neo-Kantianism, in Phenomenology, in 'Existenz-philosophie' and in some forms of Neo-Vitalism" on the one hand and the "extreme axiomatic tendencies" on the other. In 1959, *Synthese* announces "changes in our editorial policy" that aim at still greater multidisciplinarity, and warning that the methodology of the social sciences "should beware of the pseudoexactness resulting from a blind imitation of mathematical and physical procedures" (Esser et al., 1959). It is not before 1966, when Jaakko Hintikka takes up responsibility for *Synthese*, that the official statements of the editors provide a more determined attempt to clarify the role of logic in *Synthese*, and to make the "philosophy of science" the defining concept for *Synthese*'s program (Hintikka, 1966).

4 Formalization: From Meccano to mathematics

It has already become evident that, among the highly diversified group of thinkers that has been associated with the development of modern logic, *formalization* was not unequivocally seen as one of the most fundamental tasks. As an alternative, one could also opt for the task of giving each concept its proper place within a larger continuum of types of language. This attitude also marks the personal interest that Brouwer, absolutely the most outstanding logician in the significs' circle, took in this movement. For him, significs had to fulfil the centrally important function of clarifying the affective dimensions of language. He, too, seems to think in terms of a continuum in which these needs are linked to his logical and mathematical

---

[15] Cf. also Groot (1937).

[16] Cf. also Godefroy et al. (1939), where Driesch's "Ordnungslehre" is described as "both a real ['zakelijke'] and a formal logic" (p. 459).

endeavours.[17] Interestingly, one can find in the group of authors discussed so far attempts to give some kind of formalization even to this apparently strictly anti-formal attitude, thereby demonstrating why it is difficult to ascribe a clearly defined role to formalization.

One of the clearest confessions of faith regarding the role of logic in the early volumes of *Synthese* is to be found in 1946 in an "Introduction to an introduction": one of the greatest achievements of "the last few decades" is "a deepening of logic, not of academic logic with its fixed system of syllogisms, which has after all received a heavy blow from Brouwer's intuitionism, but logic reborn and renewed from the spirit of mathematics and with which the names of Frege and Russell may be said to be indissolubly connected". Evidently, a broad range of new discoveries was deemed relevant (Clay et al., 1946c, p. 102). But the very same text warns against falling into extremes, and argues against too extreme forms of formalization, the "disease of mathematicians", in a quote from Hermann Weyl that was possibly transmitted via a text by Schlick.[18] Formalization thus enters into a discourse that also concerns the question of whether it is legitimate to abstract from the more intuitive aspects of cognition. In the early *Synthese* a rather critical attitude prevails with respect to formalization.[19]

Particularly telling is the way in which Gerrit Mannoury (1867–1956)— an autodidact who ended up as professor of mathematics at Amsterdam— both criticized and made use of formalization.[20] In a short reply to a lecture by Donald C. Williams (1937), Mannoury takes a sharply critical stance towards formalization: "it is impossible to find a formal delimitation for the human activity called science and [...] trying to do so must be regarded as unscientific" (Mannoury, 1937a, p. 369). His argument lies in a critique, not

---

[17]Cf. Brouwer's "specific declaration" in (Brouwer, 1939, p. 9); for Brouwer, according to this document, the role of significa does not so much consist in "taalkritiek", a critical study of language, but rather in "detecting the affective elements which form the basis of the function of words", and in the "creation of a new vocabulary" that makes it possible to communicate over the spiritual tendencies of life.

[18]Interestingly, the original quote establishes a connection with yet more distant philosophers: "Wer freilich in logischen Dingen nur formalisieren, nicht sehen will—und das Formalisieren is[t] ja die Mathematikerkrankheit—wird weder bei Husserl noch bei Fichte auf seine Rechnung kommen" (Clay et al., 1946c, p. 104). The quote is from Weyl (1919); Schlick himself quotes this passage in (Schlick, 1926, p. 175).

[19]Detlefsen (2005, p. 237) reconstructs the history of formalism according to these lines: "they moved towards a conception of rigor that emphasized abstraction from rather than immersion in intuition and meaning".

[20]Gerrit Mannoury started his career as "privaatdocent" for the logical foundations in mathematics in Amsterdam and became then extraordinary professor in mathematics and later ordinarius first for "meetkunde, werktuigkunde en de wijsbegeeerte der wiskunde", then for "meetkunde, mechanica en wijsbegeerte der wiskunde" in Amsterdam.[21] For a more detailed account of the role of formalization within the significal movement, the contributions by Brouwer (1937) are of primary importance.

spelled out in this short statement, of the tenet that ideas may be "chained up to words".

What is needed is a more flexible analysis of language. Mannoury presents such an account in a programmatic paper on the significal foundations of mathematics (Mannoury, 1934), his contribution to the Paris conference on unity of science. He draws a distinction, in close parallelism with the two levels of language already mentioned, between two types of speech acts (hereby explicitly employing the terminology of "speech acts"). For him, there exist indicative, i.e., object-directed, and volitional or imperative modes of speech, and thus it becomes immediately evident that a labelling of ideas by words is based on a far too simplistic view of language. As a means for investigating language he suggests a thorough study of the development of language in children and explicitly quotes Brouwer's studies in the philosophy of language as an illustration of his own ideas. The distinguishing feature of mathematics is seen in its universal relationalism: mathematics works exclusively on the level of relations; mathematical definitions relate concepts or symbols without asking whether they correspond to any kind of reality. As a remedy, Mannoury asks for a well-defined psychological foundation without which any kind of distinction—and, one is supposed to add, any talk in terms of relations—is bound to remain "leeres Gerede", empty chatter. Mannoury here takes psychology in an extremely broad sense that again jumps over seemingly deep chasms between different traditions when citing Reichenbach, Russell, Weyl, but also Hugo Dingler, as relevant authors contributing to the kind of psychology he aims at. This extreme extension of the idea of psychology might offer the only way to bypass the rather obvious charge of psychologism; if psychology itself can be linked to the endeavour to provide a basic, fundamental discipline that lies beyond the traditional distinctions between, e.g., mathematical and empirical disciplines, the charges of psychologism lose force.[22]

Mannoury then sketches in some detail a systematical approach to language that he deems compatible with his basic view of language, but that, on the other hand, displays some typical elements of formal languages. Again, an important innovation from logic is taken up, namely the role of relations that have to supplement the elements of language or thought that are detected via introspection: "'Elements' and 'relations' are, therefore,

---

[22]In all these aspects, there are highly interesting links with the school of introspective thought psychology in Würzburg; protagonists of this school aimed at making this form of psychology into a genuine experimental science and also reflected on new forms of science (Oswald Külpe, mentioned above, is an example). Karl Bühler, also a member of the Würzburg school, did pioneering studies in child psychology that might be related to the significs' idea of investigating the language of children. On the 'Würzburger Schule', cf. Ziche (1999); Kusch (1999).

the necessary building blocks on the basis of which the whole terminology has to be developed, in precisely the way the mathematician constructs a set theoretical or combinatorial-topological terminology" (Mannoury, 1937b, p. 186–187). Mannoury's criterion for a successful reconstruction of language is the "Anschluß", the compatibility with the "living languages", and this criterion is understood as explicitly non-formal. In some respects, in particular in Mannoury's distinction between an "Es-Sprache" and an "Ich-Sprache", we can detect strong links with Carnap's investigations into a possible physicalisation of the language of psychology that were published in 1932/33, also in *Erkenntnis* (Carnap, 1933).

But what can then be the use, in a setting that is highly critical with regard to formalization, for a formalistic account of language? Mannoury himself suggests a formalization, first of the terminology of psychology, and he explicitly intends this formalization to "fill the gap, at least partially", between psychological and physical (or physiological) terminology—and that certainly is a typical task of significs, in complete agreement with the idea that there are multiple levels of language and with the anti-reductive program of integrating the various extreme positions into one over-arching theory.

His formalization can be presented in very concrete form. Obviously, he made use of a Meccano-construction kit (Mannoury had children) in order to build models for the elementary structures of psychical acts, thus reifying the "building blocks" he spoke about, and he presents these models in his lectures (the illustrations in Figures 1 and 2 come from Mannoury, 1934, p. 302).[23]

These models are based on the idea that the elementary facts of psychical life consist in two kinds of excitations or drives, namely those of seeking pleasant and of avoiding unpleasant or undesirable circumstances. These can be considered as two poles generating an activity line of associations that can then be integrated into a larger, two-dimensionally extended network. These two-dimensional representations are again to be understood as a projection of a three-dimensional network, and it is this three-dimensional structure that Mannoury presents in his lectures via a Meccano-based model. The models indeed display, in their tendency towards abstraction and symbolic representation, important aspects of formalization, and it is worth noting that one can apply to them purely mathematical techniques such as projecting them onto different planes.

---

[23]"Meccano" construction sets—originally called "Mechanics Made Easy", and thus incorporating an interesting element of popularizing science—were produced since the very first years of the 20th century; the name was adopted in 1908; cf. Love and Gamble (1986); Bowler (2009).

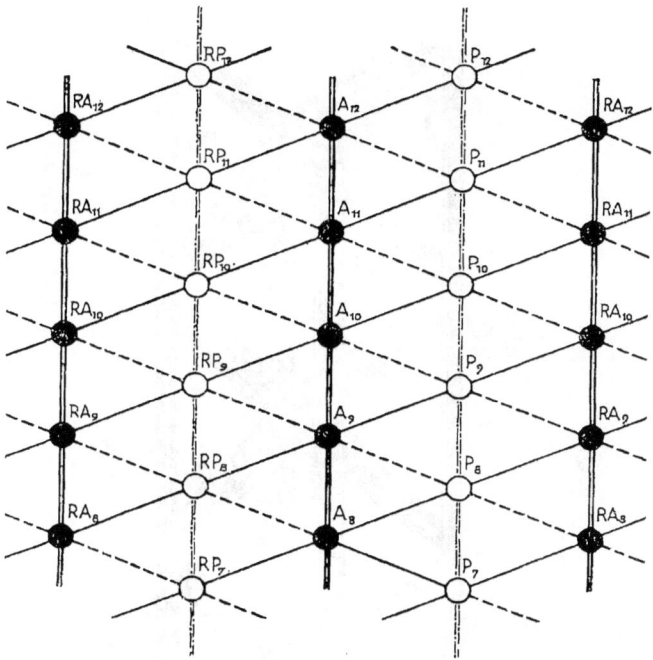

Figure 1.

Mannoury is quick to stress the restrictions of this model; it still is a "dürftiges Formalisierungssystem" (Mannoury, 1934, p. 306), a rather impoverished system of formalization that requires "further differentiation and filling in" in order to fit our ordinary language. But what, then, can be the function of this type of formalization? It is viewed as a tool that can, for instance, point towards "missing links"—this evolutionary jargon is quoted in English in the German text—, but can never serve as an ultimate foundation, "als eine Art Anfangspunkt" for a "Begriffslehre", a theory of concepts (Mannoury 1934, p. 307), from which a theory of concepts might be derived in a logical way. The foundations of language cannot rest on "dead formulae" but must be based upon the "living facticity of the concepts" ("die lebende Tatsächlichkeit der Begriffe selbst"), a phrase that with its phenomenological and humanities-based ring again evokes ideas from highly heterogeneous strands in philosophy.

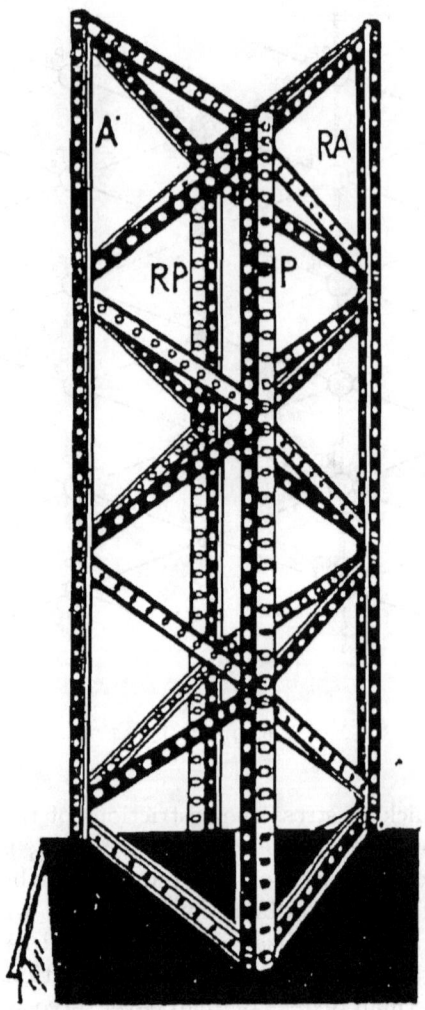

Figure 2.

5 Summary: multiple discoveries and dis-integration

Although in many respects related to the paradigm case for a multiple discovery, energy conservation, the multifarious claims regarding the discovery of logic follow a fundamentally different script. Whereas it becomes possible to integrate the various discoveries of energy conservation into one overarching framework (rooted in basic natural science with generally accepted methodological standards), this becomes impossible in the case of the alternative discoveries of logic: scientists that—from today's perspective, at least—not only appear to work in very different fields employing incommensurable methodologies but also to exemplify completely different forms of rationality, lay claim, equally, to be the discoverers of modern logic.

The conclusions one can draw are far more concrete than in the case of multiples. For the latter, the theoretical background still is in a strangely embryonic state, given the attention devoted to this phenomenon by classical authors such as Merton and Kuhn. Rather few recent approaches to this problem exist, and Merton and Kuhn themselves at best touch upon some central issues. The seemingly overwhelmingly documented fact of multiple discoveries in science lends itself to theories about the psychology of discovery, the role of genius in science, or to a phrasing in rather vague categories such as that a discovery "lies in the air" or that the "time is ripe" for such a discovery (De Solla Price, 1963, p. 66–67; Kuhn, 1959, p. 70).[24]

The scenarios of alternative discoveries do not require the problematical assumption that the various protagonists acted "in complete ignorance of each other's work" (Kuhn, 1959, p. 70). Rather, they point explicitly towards the importance of studying the interplay between distance and *unity* involved in the various alternatives, and to address the forms of distance separating the individual discoveries. Alternative claims frequently lead to a future *segregation* of the diverse fields, and in this process bring the relevant criteria for exhibiting unity versus diversity between fields into sharper focus.

The case studies presented here show that some of the most crucial concepts of modern logic, and of a philosophy that takes modern logic as its model, can be brought to bear on a surprisingly broad set of positions. Issues such as formalization and the problem of "Scheinprobleme" can be treated in both a boldly harmonizing way, embedding mathematically inspired forms of logic into a yet more comprehensive context, and in a fashion that leads to an increasingly more precise, and more thorough demarcation between scientific and philosophical fields.

The way in which comprehensiveness was to be achieved can itself be used as a criterion for giving precise contours to the various schools of thought co-existing around 1900. Two examples from the context that has been discussed here may be adduced. The writer Frederik van Eeden (who would make a particularly interesting case study for the fusion of different forms of rationality, given his interest in parapsychology), begins his book *Redekunstige grondslag van verstandhouding* ("Logical foundations of agreement")—which has been compared in form and content, to some ex-

---

[24]Lamb and Easton (1984), in their comprehensive classification of forms of multiple discoveries equally presuppose that genuine cases of multiple discovery are easy to find.

tent, with Wittgenstein's *Tractatus*—on a note reminiscent of the Vienna Circle's manifesto: there is a "need [...] in many people, for certainty, clarity and logical connections". Towards this end, he embarks on a "dialectical" clarification of language wherein the "symbolical meaning" of each word should be strictly preserved (van Eeden, 1897, p. 12*). This modifies and broadens the ideal of clarity considerably. What is more, the same holds for his view of logic: "the harmony of the parts, logical connection, the law of formulated thoughts that one calls logic, can only be reached in the case of a simple structure", beyond that we have to sacrifice the "harmony of living things" and the "harmony of reason" (van Eeden, 1897, p. 12*).

The dialectics of differentiation and integration become particularly clear in the division of labour suggested by the Amsterdam logician Evert W. Beth in a discussion of the meaning of significs for logic: "logical methods are adequate with regard to mathematical reasoning"; by contrast, the application of significal methods "will be efficient in a study of propaganda" (Beth, 1948, p. 84). But these two attitudes need not conflict: "both of them aim at clarifying our terminology, at improving our means of understanding". Plus, "in intermediate domains—in experimental science, in historiography, in literary criticism—, an intermediate point of view and a simultaneous application of both logical and significal methods will be most appropriate. Logical empiricism, for instance, may be considered as an intermediate position adapted to the peculiar nature of experimental science." What makes this bold harmonization particularly interesting is the juxtaposition of experimental science with the prototypical humanities; Beth, evidently, does not see a privileged link between mathematics, logic and experimental science, nor does he take experimental science, qua its ideal of rationality, to stand any closer to logic than to the humanities.[25]

It also becomes obvious where the fault lines run along which these demarcation processes take place. Core concepts such as "science" and "formalization" are by no means clearly defined. Witness the juxtaposition of "experimental science" and the humanities in Beth and comparable statements in the writings of Ostwald. Consequently, such concepts cannot serve as guidelines along which to structure the development of logic, being themselves subject to revision in these processes. Mannoury tackles these issues by situating logic in a broad theoretical and cultural context and, thereby, realizes the significs' demand for integration of affective and logical dimensions. His inaugural lecture entitled *On the social meaning of the mathematical form of thinking* first restricts the role of formalization (Mannoury, 1917). He argues that formalization in mathematics shares its function with language in general: the ultimate goal lies in making thoughts and ideas

---

[25]Cf. also the strong interest in the history of logic that is to be found in, e.g., Beth (1944a,b, 1946).

available as fixed points of reference. Then, significantly, he goes a step further and contrasts two conceptions of mathematics. On the one hand, mathematics can be seen as "gevoelloos, is onwezenlijk, [...] dood" ("without emotions, not essential, dead") adding, pensively: "En toch, en toch..." ("But still, but still [...]"). On the other hand, he stresses that contemporary debates on the foundations of mathematics were directly linked to deeper needs. These statements retrace the interpenetration of various lines of traditions: Beth extolls discoveries that stood at the cradle of modern logic (multidimensional geometry, transfinite numbers, point sets) in a hymnal, metaphysically charged language: these are the "sources... from which flows forth the clear certainty that enlightens the path of mathematics" and "sparks of light kindled by the last century in the field of mathematics".

Bibliography

Beth, E. W. (1944a). *De wijsbegeerte der wiskunde van Parmenides tot Bolzano*. Standaard, Antwerpen.

Beth, E. W. (1944b). *Geschiedenis der logica*. Servire, Den Haag.

Beth, E. W. (1946). Historical studies in traditional philosophy. *Synthese*, 5:258–270.

Beth, E. W. (1948). Significs and logic. In *Feestbundel aangeboden door vrienden en leerlingen aan Prof. Dr. H. J. Pos*, pages 86–95, Amsterdam. Noord-Hollandsche uitgevers.

Bowler, P. J. (2009). *Science for all. The popularization of science in early twentieth-century Britain*. University of Chicago Press, Chicago, IL.

Bowne, G. (1966). *The philosophy of logic 1880–1908*. Mouton & Co, London.

Brouwer, L. E. J. (1937). Signifiese dialogen. *Synthese*, 2:168–174, 261–268, 316–324.

Brouwer, L. E. J. (1939). *Signifische dialogen*. Bijleveld, Utrecht.

Carnap, R. (1928). *Der logische Aufbau der Welt*. Weltkreis-Verlag, Berlin-Schlachtensee. Page numbers refer to the English translation (Carnap, 2003).

Carnap, R. (1933). Psychologie in physikalischer Sprache. *Erkenntnis*, 3:107–142.

Carnap, R. (2003). *The logical structure of the world and pseudoproblems in philosophy*. Open Court, Chicago and La Salle, IL. Translated by Rolf A. George.

Carnap, R., Hahn, H., and Neurath, O. (1929). Wissenschaftliche Weltauffassung. Der Wiener Kreis. In Haller, R. and Rutte, H., editors, *Otto Neurath: Gesammelte philosophische und methodologische Schriften. Part 1*, pages 299–336, Wien. Hölder-Pichler-Tempsky. 1981.

Clay, J., Godefroy, J., Kruseman, W., Vuysje, D., Westendorp Boerma, N., and Westerman Holstijn, A. (1946a). The course of Synthese. Translation of an article issued Dec. '39. *Synthese*, 5:21–24. Signed "The Editors".

Clay, J., Godefroy, J., Kruseman, W., Vuysje, D., Westendorp Boerma, N., and Westerman Holstijn, A. (1946b). Introduction 1946 (English version). *Synthese*, 5:9–11. Signed "The Editors".

Clay, J., Godefroy, J., Kruseman, W., Vuysje, D., Westendorp Boerma, N., and Westerman Holstijn, A. (1946c). Introduction to an introduction. *Synthese*, 5:102–104. Signed "The Editors".

de Jager, H. (1992). *Het beeldende denken. Leven en werk van Mathieu Schoenmaekers*. Ambo, Baarn.

De Solla Price, D. J. (1963). *Little science, big science*. Columbia University Press, New York.

Detlefsen, M. (2005). Formalism. In Shapiro, S., editor, *Philosophy of mathematics and logic*, pages 236–317, Oxford. Oxford University Press.

Driesch, H. (1909). *Philosophie des Organischen. Gifford-Vorlesungen gehalten an der Universität Aberdeen in den Jahren 1907–1908*. Wilhelm Engelmann, Leipzig.

Driesch, H. (1913). *Die Logik als Aufgabe. Eine Studie über die Beziehung zwischen Phänomenologie und Logik zugleich eine Einleitung in die Ordnungslehre*. J.C.B. Mohr (Paul Siebeck), Tübingen.

Driesch, H. (1923). *Ordnungslehre. Ein System des nichtmetaphysischen Teiles der Philosophie*. New ed. Jena, Diederichs.

Driesch, H. (1924). *Biologische Probleme höherer Ordnung*. Barth, Leipzig.

Driesch, H. (1926). *Metaphysik der Natur*. Handbuch der Philosophie. Oldenbourg, München.

Eley, L., editor (1970). *Philosophie der Arithmetik. Mit ergänzenden Texten (1890-1901)*, volume 12 of *Husserliana*. Nijhoff, Den Haag.

Esser, P., Frank, P., Groenewold, H., Hofstra, S., Kazemier, B., Kruseman, W., Naess, A., Piaget, J., Raven, C., Woodger, J., and Vuysje, D. (1959). Introduction to the new volume. *Synthese*, 11:3–4. Signed "The Editors".

Godefroy, J., Groot, H., de Hartog, A., van Hinloopen Labberton, D., Kruseman, W., Schoenmaekers, M., Westendorp Boerma, N., and Westerman Holstijn, A. (1936a). De synthetische gedachte in het buitenland. *Synthese*, 1:106–112. Signed "Redactie".

Godefroy, J., Groot, H., de Hartog, A., van Hinloopen Labberton, D., Kruseman, W., Schoenmaekers, M., Westendorp Boerma, N., and Westerman Holstijn, A. (1936b). Inleiding. *Synthese*, 1:1–6. Signed "Redactie".

Godefroy, J., Groot, H., de Hartog, A., van Hinloopen Labberton, D., Kruseman, W., Schoenmaekers, M., Westendorp Boerma, N., and Westerman Holstijn, A. (1936c). Moritz Schlick †. *Synthese*, 1:193–201. Signed "Redactie".

Godefroy, J., Groot, H., de Hartog, A., van Hinloopen Labberton, D., Kruseman, W., Schoenmaekers, M., Westendorp Boerma, N., and Westerman Holstijn, A. (1936d). Significa. De "Wiener Kreis" en de eenheid der wetenschap. Logische analyse. De Hollandsche significi. De taak, die "Synthese" zich stelt. *Synthese*, 1:325–339. Signed "Redactie".

Godefroy, J., Groot, H., de Hartog, A., van Hinloopen Labberton, D., Kruseman, W., Schoenmaekers, M., Westendorp Boerma, N., and Westerman Holstijn, A. (1939). Hans Driesch als wijsgeer. De kentheoreticus en de naturphilosoof. *Synthese*, 4:455–466. Signed "Redactie".

Görs, B., Psarros, N., and Ziche, P., editors (2005). *Wilhelm Ostwald at the crossroads of chemistry, philosophy and media culture*. Universitätsverlag, Leipzig.

Groot, H. (1937). Metafysika en logistiek als doorgangsstadia. *Synthese*, 2:486–491.

Haaparanta, L., editor (2009). *The development of modern logic*. Oxford University Press, Oxford.

Hankel, H. (1867). *Theorie der complexen Zahlensysteme insbesondere der gemeinen imaginären Zahlen und der Hamiltonschen Quaternionen nebst ihrer geometrischen Darstellung*. Voss, Leipzig.

Hansen, F.-P. (2000). *Geschichte der Logik des 19. Jahrhunderts. Eine kritische Einführung in die Anfänge der Erkenntnis- und Wissenschaftstheorie.* Königshausen & Neumann, Würzburg.

Hegselmann, R. and Siegwart, G. (1991). Zur Geschichte der 'Erkenntnis'. *Erkenntnis*, 35(461–471).

Heijerman, A. and van den Hoven, M. (1986). *Filosofie in Nederland. De Internationale School voor Wijsbegeerte als ontmoetingsplaats 1916–1986.* Boom, Meppel.

Hintikka, J. (1966). Editorial. *Synthese*, 16:1–3.

Husserl, E. (1901). Das Imaginäre in der Mathematik. Lecture notes. Page numbers refer to the published version in Eley (1970, pp. 430–444).

Ihmig, K.-N. (1997). *Cassirers Invariantentheorie der Erfahrung und seine Rezeption des 'Erlanger Programms'.* Meiner, Hamburg.

Jahn, I. (1998). *Geschichte der Biologie. Theorien, Methoden, Institutionen, Kurzbiographien.* Fischer, Jena.

Kuhn, T. S. (1959). Energy conservation as an example of simultaneous discovery. In Clagett, M., editor, *Critical problems in the history of science*, pages 321–356, Madison WI. University of Wisconsin Press. Page numbers refer to Kuhn (1977, pp. 66–104).

Kuhn, T. S. (1977). *The essential tension. Selected studies in scientific tradition and change.* Chicago University Press, Chicago, IL.

Kusch, M. (1999). *Psychological knowledge: A social history and philosophy.* Routledge, London.

Lamb, D. and Easton, S. M. (1984). *Multiple discovery. The pattern of scientific progress.* Avebury, Letchworth, Herts.

Love, B. and Gamble, J. (1986). *The Meccano system and the special purpose Meccano sets, 1901–1979.* New Cavendish Books, London.

Mannoury, G. (1917). *Over de sociale betekenis van de wiskundige denkvorm.* Noordhoff, Groningen.

Mannoury, G. (1934). Die signifischen Grundlagen der Mathematik. *Erkenntnis*, 4:288–309, 317–345.

Mannoury, G. (1937a). Schlußbemerkung. *Erkenntnis*, 7:369.

Mannoury, G. (1937b). Signifische Analyse der Willenssprache als Grundlage einer physikalistischen Sprachsynthese. *Erkenntnis*, 7:180–188.

Merrill, D. D. (1990). *Augustus de Morgan and the logic of relations.* Kluwer, Dordrecht.

Merton, R. K. (1961). Singletons and multiples in science: a chapter in the sociology of science. *Proceedings of the American Philosophical Society*, 105(5):470–486.

Merton, R. K. (1963). Resistance to the systematic study of multiple discoveries in science. *European Journal of Sociology*, 4:237–249.

Natorp, P. (1923). *Die logischen Grundlagen der exakten Wissenschaften.* Teubner, Leipzig.

Neurath, O. (1946). After six years. *Synthese*, 5:77–82.

Ostwald, W. (1909). Das System der Wissenschaften. *Annalen der Naturphilosophie*, 8:266–272.

Ostwald, W. (1914). *Moderne Naturphilosophie. I. Die Ordnungswissenschaften.* Akademische Verlagsanstalt, Leipzig.

Peckhaus, V. (1997). Review of Jarmo Pulkkinen, *The threat of logical mathematism. A study on the critique of mathematical logic in Germany at the turn of the 20th century*, Peter Lang: Frankfurt a.M. 1994. *History and Philosophy of Logic*, 18:115–120. Essay review of Pulkkinen (1994).

Pulkkinen, J. (1994). *The threat of logical mathematicians. A study on the critique of mathematical logic in Germany at the turn of the 20th century.* Peter Lang, Frankfurt a.M.

Pulkkinen, J. (2005). *Thought and logic. The debates between German-speaking philosophers and symbolic logicians at the turn of the 20th century.* Peter Lang, Frankfurt a.M.

Schlick, M. (1926). Erleben, Erkennen, Metaphysik. *Kant-Studien*, 31:146–158. Page numbers refer to (Stöltzner and Uebel, 2006, pp. 169–186).

Schlick, M. and Hertz, P., editors (1921). *Hermann von Helmholtz, Schriften zur Erkenntnistheorie.* Springer, Wien.

Schlimm, D. (2011). On the creative role of axiomatics. The discovery of lattices by Schröder, Dedekind, Birkhoff, and others. *Synthese*, 183(1):47–68.

Schmitz, H. W. (1990). *De Hollandse Significa. Een reconstructie van de geschiedenis van 1892 tot 1926*. Van Gorcum, Assen.

Schoenmaekers, M. (1936). Zinzeggende beeldspraak. *Synthese*, 1:7–10, 34–39.

Sluga, H. (1997). Frege on meaning. In Glock, H.-J., editor, *The rise of analytic philosophy*, pages 17–34, Oxford. Blackwell.

Stöltzner, M. and Uebel, T. E., editors (2006). *Wiener Kreis: Texte zur wissenschaftlichen Weltauffassung von Rudolf Carnap, Otto Neurath, Moritz Schlick, Philipp Frank, Hans Hahn, Karl Menger, Edgar Zilsel und Gustav Bergmann*, volume 577 of *Philosophische Bibliothek*. Meiner, Hamburg.

van Dalen, D. (1999). *Mystic, geometer, and intuitionist. The life of L.E.J. Brouwer. Part 1*. Clarendon, Oxford.

van Dantzig, D. (1956). Mannoury's impact on philosophy and significs. *Synthese*, 10:423–431.

van Eeden, F. (1897). *Redekunstig grondslag van verstandhouding*, volume 3 of *Studies*. Versluys. Page numbers refer to Vieregge et al. (2005).

Vieregge, W. H., Schmitz, H. W., and Noordegraaf, J., editors (2005). *Frederik van Eeden. Logische Grundlage der Verständigung / Redekunstig grondslag van verstandhouding*, volume 127 of *Germanistik ZDL-Beiheft*. Steiner, Stuttgart.

Vollenhoven, D. (1921). Einiges über die Logik in dem Vitalismus von Driesch. *Biologisches Zentralblatt*, 41:337–358.

von Helmholtz, H. (1878). *Die Thatsachen in der Wahrnehmung*. August Hirschwald, Berlin. Page numbers refer to Schlick and Hertz (1921, pp. 59–78).

Weyl, H. (1919). Der circulus vitiosus in der heutigen Begründung der Analysis. *Jahresberichte der deutschen Mathematiker-Vereinigung*, 28:85–92.

Whitehead, A. N. (1898). *A treatise on universal algebra with applications*. Cambridge University Press, Cambridge.

Williams, D. C. (1937). The realistic interpretation of scientific sentences. *Erkenntnis*, 7:169–178.

Ziche, P., editor (1999). *Introspektion. Texte zur Selbstwahrnehmung des Ichs*. Springer, Wien.

Ziche, P. (2006). "Wissen" und "hohe Gedanken". Allgemeinheit und die Metareflexion des Wissenschaftssystems im 19. Jahrhundert. In Hagner, M. and Laubichler, M. D., editors, *Der Hochsitz des Wissens. Das Allgemeine als wissenschaftlicher Wert*, pages 129–151, Zürich. Diaphanes.

Ziche, P. (2008). *Wissenschaftslandschaften um 1900. Philosophie, die Wissenschaften und der 'nicht-reduktive Szientismus'*. Chronos, Zürich.

Ziche, P. (2009). Wilhelm Ostwald als Begründer der modernen Logik. Logik und künstliche Sprachen bei Ostwald und Louis Couturat. In Pirmin, S.-W., Heiner, K., and Nikos, P., editors, *Ein Netz der Wissenschaften? Wilhelm Ostwalds "Annalen der Naturphilosophie" und die Durchsetzung wissenschaftlicher Paradigmen*, pages 46–66, Leipzig. Sächsische Akademie der Wissenschaften.

www.ingramcontent.com/pod-product-compliance
Lightning Source LLC
Chambersburg PA
CBHW071425150426
43191CB00008B/1043